# THE *Scented Garden*

THE

# Scented Garden

A COMPLETE GUIDE TO GROWING AND USING FRAGRANT PLANTS

Pamela Allardice

Angus&Robertson
An imprint of HarperCollins*Publishers*

For Greg, despite remarking bitterly that digging, pruning, drilling drainage holes, composting and mowing were not part of his weekend plans; and for Edward, who thinks weeds are 'pitty', but methodically pulls up all the new seedlings. With all my love.

*AN ANGUS & ROBERTSON BOOK*
*An imprint of HarperCollinsPublishers*

*First published in Australia in 1992 by*
*CollinsAngus&Robertson Publishers Pty Limited (ACN 009 913 517)*
*A division of HarperCollinsPublishers (Australia) Pty Limited*
*25–31 Ryde Road, Pymble NSW 2073, Australia*

*HarperCollinsPublishers (New Zealand) Limited*
*31 View Road, Glenfield, Auckland 10, New Zealand*

*HarperCollinsPublishers Limited*
*77– 85 Fulham Palace Road, London W6 8JB, United Kingdom*

*Distributed in the United States of America by*
*HarperCollinsPublishers*
*10 East 53rd Street, New York NY 10022, USA*



*National Library of Australia*
*Cataloguing-in-Publication data:*

*Allardice, Pamela, 1958-*
*The scented garden.*

*Bibliography.*
*Includes index.*
*ISBN 0 207 17104 1.*

*1.Gardens, Fragrant. 2. Aromatic plants. I. Title.*
*635.968*

*Cover: gardenia inset, Lorna Rose; background, International Photographic Library.*
*Typeset by Midland Typesetters Pty Ltd, Victoria.*
*Printed in Hong Kong.*

*5 4 3 2 1*
*95 94 93 92*

*Title page: the beautiful and fragrant magnolia blossom.*

*Contents page:* Crocus *Sp.*

# Contents

THE ROMANTIC STORY OF

# Scented Plants

The appeal of scented flowers and leaves, and their many byproducts, is as irresistible today as it was in that first fragrant plot, the Garden of Eden. Plants have been used to flavour food and drink, scent clothes and uplift the spirits for many centuries.

◂ The dewy fragrance of a scented garden is spiced with the chill of the morning mist.

The first scented plant referred to in the Bible was *bdellium*, the Greek name for the prickly shrub *Balsamodendron myrrha*. Bdellium provided an aromatic gum from which myrrh was made, and grew in the region between the Tigris and the Euphrates known as Babylonia (Eden was thought to have been situated at the junction of these two mighty rivers.) This was and is an intensely hot and dry area and has long been regarded as the source of the most potent scented timbers and leaves. Dionysius wrote that '. . . in fortunate Arabia, you can always smell the sweet perfume of marvellous spices, whether it be incense or myrrh.' Early people valued these aromatic spices and gums very highly, not only for their ability to preserve or flavor food, but also for their cooling properties. In *Herbal of the Bible* (1597), Newton recorded that Palestinians hung garlands of fresh fragrant boughs about their rooms and strewed green leaves on their beds to reduce the effect of the heat.

Perfume was integral to ancient Egyptian civilization. According to Genesis, aromatic plants were first imported by Ishmaelite traders who traveled from Gilead with their camels, bearing spices and balm and myrrh. This 'balm' is believed to have been the exudation of *Pistacia lentiscus*. Also known as 'mastic', this plant was used both as an incense for religious rites and as a chewy pastille, much favored by children.

Other highly prized aromatic plants included spikenard (*Nardostachys jatamansi*), a member of

### Cargoes

*Quinquereme of Nineveh from distant Ophir*
*Rowing home to haven in sunny Palestine,*
*With a cargo of ivory*
*And apes and peacocks,*
*Sandalwood, and cedarwood, and sweet white wine*

John Masefield, *The Sea Poems*

the valerian family, whose scented roots were obtained from the Himalayas, and frankincense, which was obtained from the bark of *Boswellia serrata.* Frankincense was widely used as incense, but was most expensive, for the tree grew only on rocky hillsides and ravines, making collection and transportation very difficult. It was as valued as gold and when the three kings Caspar, Melchior and Balthazar came to pay homage to the baby Jesus, they brought gold, frankincense and myrrh as gifts. Frankincense has primarily been regarded as a holy plant and featured in the religious rituals of all ancient nations, particularly the Israelites. The instructions Moses received from God regarding frankincense were detailed carefully:

*And thou shalt make an altar to burn incense upon, and thou shalt overlay it with pure gold . . . And Aaron shall burn thereon sweet incense every morning. When he dresseth the lamps he shall burn incense upon it. And, at even, he shall burn incense upon it.*

Oil of cedarwood from the dense forests of Mount Lebanon was also much in demand by the Egyptians, both for religious and funereal rites and for fragrant cosmetics and ointments. They believed cedarwood was imperishable and could preserve life; it was used for building coffins, ships and temples and burnt as an offering to the gods. The builders of antiquity had vast requirements for cedarwood, and forests were eventually exhausted. One of the main offenders was King Solomon who felled more than 500 trees when building his great temple in Jerusalem. Cedarwood was also used to construct the famous temple to Diana at Ephesus. Built in the sixth century BC, Roy Genders tells us that the temple was one of the Seven Wonders of the World—it spread over an acre and its columns were massive cedar trunks more than 55 feet high and 6 feet around. According to legend, this mighty place was burned to the ground the night Alexander the Great was born, 200 years later.

### The French Queen's Perfume

*First burn chips of Cypress in the Chamber a pretty while, the doors and windows being shut. Then take Damask Rose water a pinte: white Sugar Candy an ounce: Put them in a perfuming pan and let them boyl softly on the embers.*

Wm Salmon, *Polygraphics,* seventeenth century

For several thousand years the 'Incense Road' between Arabia and Egypt was traveled by the wealthy spice caravanserai. Trade in incense was very risky and often whole camel trains were hijacked by marauding tribesmen. Losses, in both men and goods, were high on the long journey, but the rewards for a successful merchant were correspondingly great.

The entire economy of the small, glamorous kingdom of Sheba in the south of Africa depended on the sale of frankincense and myrrh. In the tenth century BC when the expanding empire of King Solomon threatened to close the Incense Road, the Queen of Sheba decided to be her own ambassador and set off by camel train on the arduous journey to Jerusalem where Solomon lived with his 700 wives and 300 concubines. Undaunted even by this amount of competition, the Queen overwhelmed the bemused Solomon with flattery and gifts. She presented him with 31 incense trees, a cargo of ebony, sandalwood, ivory, gold, jewels, apes and, ultimately, herself. Solomon succumbed to her wiles and the Queen returned to Sheba smugly triumphant. Not only had she secured the freedom of the Incense Road, she had also won the contract to supply Solomon's empire with frankincense.

The Egyptians made extensive use of fragrant plants and spices in their religious ceremonies, in the embalming of their dead and for adorning their own bodies. Scented timbers were first used as offerings for the gods and burnt on altars in temples throughout Egypt. The first manufactured perfumes were, therefore, a product of fire. The memory of these first scents is preserved in the word 'perfume', which comes from the Latin *per* meaning 'through' and *fumus* meaning 'smoke'.

In early Egypt, every temple was filled with fragrant smoke—a mystical and sacred messenger of prayers to the heavens, spiraling upwards, never to return. However, it did not take the Egyptians long to realize that fire was not necessary to make perfume. Nor was perfume regarded as the sole property of the gods or priests for long; soon perfumed waters and oils were part of every level of the social structure. The Greek biographer Plutarch made special note of the incense burnt thrice daily in honor of the Egyptian sun god, Ra. Known as *kyphi,* it was a mixture of myrrh, henna, cinnamon, juniper, honey and raisins, all of which were steeped in sweet wine, then beaten to a paste and allowed to solidify. Kyphi was also burned customarily in Egyptian homes at night to induce sleep and ensure good dreams. Plutarch added poetically that it was made of 'those things that delight most in night'.

*It is also well to boil the flowers and leaves in water and to wash yourself therewith every morning . . .*

The Physicians of Myddvai

Women of ancient Egypt were highly scented; oil of cinnamon from Nepal was combined with honey, myrrh and almonds to make a perfume for feet and legs; it was also swallowed to sweeten the breath. Nutmeg, sweet rushes, laurel and juniper were all popular perfume ingredients. The animal smells of musk and civet, both obtained from Ethiopia, were also probably used. Such ointments were at first prepared and sold by the priests who were privy to the elaborate techniques of using aromatic plant gums in mummification. They found perfume manufacture a lucrative sideline and were, as far as we know, the earliest perfume retailers. The priests perfected the art of floral extraction, and temple reliefs show perfume being extracted from white madonna lilies (*Lilium candidum*) and the sacred blue lotus (*Nymphaea coerulea*). The lotus, or 'Cradle of the Sun', was

*Nymphaea* x *odorata*

regarded as a sacred plant. The Egyptians believed the sun had risen to the sky from the centre of a lotus flower. They saw the opening and closing of the lotus' petals with the sun's passage as a symbol of reincarnation and immortality. Tomb paintings depict both the living and the dead inhaling the flower's scent.

Every part of the henna plant (*Lawsonia inermis*) is scented, particularly its delicate white flowers, which produce a green-tinted, heavy perfume much used in perfumery, known as cyprinum. Cleopatra chose cyprinum as the scent with which to drench the sails and hangings on her barge as she prepared to meet Antony. The canopy she sat beneath was garlanded with roses and slaves puffed incense towards Cleopatra and her guests from bronze burners. Shakespeare described the scene:

*Purple were the sails and so perfumed, that*
*the winds were love-sick with them . . .*

*Cassia fistula*

At Cleopatra's banquet in honor of Antony, the whole floor was carpeted with scented roses to the depth of several feet, held in position by nets fixed to the walls. The air was heavy with incense and on arrival each guest was crowned with a chaplet of white Madonna lilies, saffron crocus and lotus flowers. To impress Antony with the power and wealth of Egypt and the suitability of herself as his consort, Cleopatra dissolved a large pearl in vinegar and drank it—thus neatly making the point that a priceless gem in his country was merely an expendable trinket in hers.

The Egyptians were most lavish with perfume in their important religious rituals, notably the embalming of the bodies of the dead. The mummification industry used most of the aromatics brought into the country and supported a huge labor force of morticians and craftspeople. After the heart, liver, lungs, and intestines were cut out, oil of cedar was injected into the body along with cassia and myrrh. The removed organs were placed in a special jar with scented oils and the body was left to dehydrate in a bath of natron. After 70 days it was wrapped in linen bandages impregnated with aromatic ointments and placed in a fragrant cedarwood coffin.

*You must in rose-time make choice of such roses as are neither in the bud, nor full blowne which you must specially cull and chuse from the rest, then take sand and dry it thoroughly well, and having shallow boxes, make first an even lay of sand, upon which you lay your rose-leaves one by one. Set this box in some warme, sunny place in a hot sunny day (and commonly in two hot sunny dayes they will be thorow dry), and thus you may have rose-leaves and other flowers to lay about your basons and windows all winter long.*

Sir Hugh Platt, *Delightes for Ladies*, 1594

The consumption of perfumes in Egypt set an example to the whole of the ancient world, particularly the Jews. When Moses led the Israelites back from Egypt to their own country, they brought many seeds of fragrant shrubs and trees and the knowledge necessary for their cultivation. Notably plentiful was the amyris shrub (*Commiphora opobalsamum*), which yields the famous Balm of Gilead. According to Exodus, Moses was given special instructions on his return to prepare a holy anointing oil:

*Thou shalt take unto thee principal spices of pure myrrh five hundred shekels; and of sweet cinnamon half so much, two hundred and fifty shekels; and of sweet calamus two hundred and fifty shekels; and of cassis five hundred shekels; and of olive oil an hin [a gallon]. And thou shalt make it a perfume . . . pure and holy.*

Moses decreed severe penalties against anyone using holy oils for private purposes. However, the Jewish women began to use perfume as a cosmetic just as their tutors, the Egyptians, had done. There are many historical references to their use, for example, the beautiful Esther underwent a year

*Rosa banksiae* 'Lutea'

of purification, '. . . to wit, six months with oil of myrrh and six months with sweet odours', before being presented to King Xerxes as his new wife.

There are also many references to the aromatic herbs and spices used by the Jewish people. In the *Song of Solomon* it is written that:

> *While the king sitteth at his table, my spikenard sendeth forth the smell thereof. A bundle of myrrh is my beloved to me; he shall lie all night betwixt my breasts. My beloved is as to me as a cluster of camphire in the vineyards of Engedi. Who is this that cometh out of the wilderness like pillars of smoke, perfumed with myrrh and frankincense with all the powders of the merchant? Thy lips, O my spouse, drop as the honeycomb; and the smell of thy garments is like the smell of Lebanon. Thy plants are an orchard of spikenard, spikenard and saffron; calamus and cinnamon; with all trees of frankincense, myrrh and aloes, with all the chief spices. I rose up to open to my beloved; and my hands dripped with myrrh, and my fingers with sweet smelling myrrh, upon the handles of the lock. His cheeks are as a bed of spices, as sweet flowers; his lips are like lilies, dropping sweet-smelling myrrh. His countenance is as Lebanon, excellent as the cedars.*

Another famed fragrant plant from early times was the Cretan Rock Rose (*Cistus ladaniferus*), or Rose of Sharon. This produced a gummy exudation known as labdanum, which was the base of many perfumed cosmetics and oils. In Syria and Crete it is still combed from the fleeces of mountain sheep that have been grazing by the plant, the same way it was harvested in Solomon's time.

*You may take of Rose leaves four ounces, cloves one ounce, lignum Rhodium two ounces, Storax one ounce and a halfe, Muske and Civet of each ten grains; beat and incorporate them well together.*

*Ram's Little Dodoen*, 1606

According to the Koran, Paradise is peopled with pretty nymphs scented with musk. Mohammed was particularly fond of musky scents and decreed that musk should be mixed with the mortar used for building his temples. He also used scented henna to dye his beard and Mohammedan women emulated his habit by rubbing henna leaves on their cheeks and hands to give them a rouged look. Mohammed decreed the henna plant to be '. . . the chief of sweet-scented flowers of this world and the next'. He also likened the 'excellence of the violet to the excellence of El-Islam above all other religions.'

Eastern gardens were planted with myrtle, jonquils and damask roses, the most famous being the 'hanging' gardens of Babylon, built by King Nebuchadnezzar for his wife, Amytes. In truth, these were not 'hanging' gardens, but roof gardens laid out on balconies. Henna is thought to have been one of the hedging plants used in these gardens and it can still be seen forming windbreaks around Middle Eastern vineyards, a function it has performed for many centuries. Records state all the plants in the gardens were aromatic and designed to please the sense of smell as much as that of sight: '. . . rich fruits o'erhang the sloping vales, and odorous shrubs entwined their undulating branches.'

The highly perfumed and heavily made-up women of Assyria were famed throughout the Middle East for their beauty. As Assyria was situated on the main part of the Incense Road, Assyrian women had the pick of exotic fragrant plants and cosmetics at their disposal. Herodotus was intrigued by the complicated beauty regimen of these women, notably their use of pumice stone to smooth the skin and their preparation of a complicated mask designed to impregnate the skin with scented oil. To prepare the latter, they would 'bruise with a stone the wood of the cypress, cedar and frankincense and pour water upon it until it became of the desired consistency. With this they anointed the body and face to impart a most agreeable odour.' This was left on the skin for at least a day before being removed.

Up until this time most perfumes had been oils or wines scented with freshly gathered flowers or bark gums. The Persians, however, were the first to refine the art of preserving fragrance. They learned how to pack the flower petals or spices in earthenware jars sealed with clay and dry them in the sun. Bowls of these dry mixes would be placed in rooms to freshen them or burned as part of a ritual fumigation. The Persians especially loved red roses, and they told the tale that a rose bush was merely a thorn bush until the sweat of the prophet Mohammed dropped onto it, after which it bore lush flowers. The Persian poet Sheikh Sadi wrote of it:

*And thou, then, musk of ambergris, I said;*
*That by thy scent my soul is ravished.*
*'Not so,' it answered; 'Worthless earth was I,*
*But long I kept the roses company;*
*Thus near its perfect fragrance to me come,*
*Else I am but earth, the worthless and the same.'*

Fragrant plants played an integral part in the lives of Indian people. Sacrifices of scented flowers and barks, notably the sacred rusa (*Andropogon nardus*) were made to Hindu deities. Lemongrass (*Andropogon citratus*), a species native to India that is used to make attar for perfumery and cosmetics, was also used to make holy anointing oils. A custom still seen today is the weaving of dried roots of vetivert (*Vetivert zizanoides*) into screens, or 'tatties', for use as shades for windows

▶ 'Softly on there every hour
Secret herbs their spices shower'.

or doors. When these are sprayed with water the violet-like fragrance is released, helping to cool as well as freshen the air. Vetivert was a particular favorite of Georgian and Victorian dandies in England; they used it as a handkerchief perfume.

At wealthy Hindu marriage ceremonies, the bride and groom stand beneath a silk canopy by a sacred fire of sandalwood (*Santalum album*). On the last day of the year, it is traditional for Indian women to sprinkle passers-by with rosewater and sandalwood essence, which are thought to wash away the sins or bad luck of that year. During religious festivals for the god Vishnu, alms are collected in exchange for sticks of sandalwood. In Tibet, sandalwood remains the most sacred of scented plants. Portable burners of gilded copper inlaid with precious stones are used daily to keep away evil spirits. Priests place sandalwood incense at the feet of the dead so that the soul may waft

To Make Oyntment of Roses

*Take oyl of Roses four ounces, white wax one ounce, melt them together over seething water, then chafe them together with Rose-water and a little white Vinegar.*

John Partridge, *The Treasurie of Hidden Secrets and Commodious Conceits*, 1586

upwards to heaven on the perfume. Dense forests of sandalwood once thrived throughout India; however, so much sandalwood has been harvested for building temples and religious artefacts (due to its religious significance and its powerful ability to repel white ants) that the number of forests has sadly dwindled.

Tulsi, or Holy Basil (*Ocimum sanctum*) is also valued in India, along with the otto derived from the nettle-like *Pogostemon patchouli*, which is native to Bengal. Patchouli is the most powerful of all plant-derived scents. When it is undiluted it is, to quote nineteenth-century perfumer Charles Piesse, '. . . far from agreeable, having a kind of mossy or musty odour.' This eastern fragrance nonetheless created a fashionable stir in Victorian England when, due to its moth-repelling properties, the crushed leaves were packed among Indian cashmere shawls imported to the country (a fashion started by Queen Victoria herself). British manufacturers tried to cash in on the popularity of these shawls but customers learned to distinguish the imitations by a quick sniff-test. If a whiff of patchouli was present, the shawl was readily accepted.

In India, roses and jasmine are plentiful. In fact, the attar of roses from Ghazipur in the northwest India was judged the best in the world until the mid-nineteenth century when the French rose fields of Grasse overtook it. The Hindi name for jasmine translates as 'moonlight of the groves' and *sambac*, made from the flowers of *Jasminum sambac*, remains a popular fragrance among young Indian girls. Kama, the Indian god of love, carries a quiver of five arrows, each specifically affecting one of the five senses. The arrow thought to enamor the sense of smell is tipped with a jasmine flower. Jasmine's romantic associations were also appreciated by the English, and Tudor poet Edmund Spenser used them as a simile in a sonnet dedicated to his wife:

*Her breasts, like lilies, 'ere their leaves be shed;*
*Her nipples, like young blossomed jessamines;*
*Such fragrant flowers do give most odorous smell*
*But her sweet odour did them all excel.*

The Chinese loved scented flowers, especially peach and magnolia blossoms, jonquils and jasmine. Confucius recorded that on all festive occasions, especially New Year's Day, houseboats and temples were lavishly decked with scented flowers and candle-lit lanterns. One game popular among well-to-do Oriental women was to 'guess the scent'. Each player was given a tiny box in which a miniature charcoal fire was lit. Different incenses, including several made from different types of jasmine, were burned and the purpose of the game was to guess which was which.

Early writings record that *sambac* was widely cultivated for use in perfumery and in tea. Each summer, in Foochow alone, more than 3 million pounds of buds were harvested for tea-making. Unopened buds were also much in demand for women to wear in their hair in the evening; as the night progressed, the buds would slowly open under the moonlight, their scent lingering in the wearer's hair. Just as in Europe, where lavender was placed under bedding to combat an oppressive atmosphere, so jasmine was garlanded about Chinese beds.

*If thou hast wisdom, hear me, Celia,*
*Thy baths shall be the juice of July-flowers,*
*Spirit of Roses and of violets,*
*The milk of unicorns, and panther's breath*
*Gathered in bags and mixed with Cretan wines . . .*

Ben Jonson, *Volpone*

The Greeks believed scented flowers and leaves were invented by the divine denizens of Mount Olympus, specifically Venus and her troupe of nymphs. According to Homer, an overpowering scent of flowers always accompanied the Goddess of Love:

*Celestial Venus hovered o'er his head,*
*And roseate unguents heavenly fragrance shed . . .*

The secrets of extracting perfume from scented plants reached mortal women through Venus. Her nymph Oenone divulged the method to Paris who in turn taught it to that legendary beauty, Helen of Troy. Other goddesses emulated Venus' habits; witness Juno's preparation when awaiting her rival's appearance:

*Here first she bathes and around her body pours*
*Soft oils of fragrance and ambrosial showers.*

Perfume in Greece rapidly passed from the gods to the commercial domain. Many herbalists set up perfume-making equipment and began producing scents for wealthy Athenians. Some gave their names to particular scents; *megallum*, for example, was formulated by a perfumer called Megallus. This preparation contained myrrh, cinnamon and cassia (*Laurus cassia*).

The Greeks enjoyed perfumes and had a great many to choose from. Apollonius of Herophila, in his *Treatise on Perfumes*, listed some of the 'best buys' of the time:

*The iris is best at Elis and at Cyzicus; perfume from roses is most excellent at Phaselis; but that made at Naples and Capua is also good. That made from crocus (saffron) reaches the highest perfection at Soli in Cilicia and at Rhodes. Essence of spikenard is best at Tarsus; the extract of the vine at Cyprus. The best perfume from marjoram and from apples comes from Cos. From Egypt comes the best essence of cyperus . . .*

*Iris unguicularis*

A SWEET-SCENTED BATH

*Take of Roses, Citron peel, Sweet flowers, Orange flowers, Jessamy, Bays, Rosemary, Lavender, Mint, Pennyroyal, of each a sufficient quantity, boil them together gently and make a Bath to which add Oyl of Spike six drops, musk five grains, Ambergris three grains.*

John Middleton, *Five Hundred Receipts*, 1734

The Greeks would apply different scents to different parts of the body. In *The Banquet,* Philoxenes describes guests at an Athenian home being offered water mixed with the juice of lily flowers to rinse their hands, before being crowned with garlands of violets (the violet was commercially cultivated just outside Athens and became the symbol of that city. At ancient Grecian games the coveted first prize was a decorative sculpture in the form of a golden violet). Antiphanes, a writer and historian of ancient Greece, went on to record the toilet habits of a nobleman:

*He bathes*
*In a large gilded tub, and steeps his feet*
*And legs in rich Egyptian unguents.*
*His jaws and breasts he rubs with thick palm oil*
*And both his arms with extract sweet, of mint;*
*His eyebrows, and his hair, with marjoram,*
*His knees and neck with essence of ground thyme . . .*

The Greeks believed that massaging the head with scented oils would clarify thought; similarly, the poet Anacreon advised men and women to massage their breasts with fragrant flower essences and scented oils to soothe the heart. Theophrastus, a respected botanist of the time, had an unrivaled knowledge of plants. For the making of perfume he records that:

*Rose and gillyflower perfumes are made from the flowers . . . made from the leaves are those culled from myrtle and dropwort . . . From roots are made the scents named from iris spikenard and sweet marjoram . . .*

Theophrastus also mentioned Sweet Flag (*Acorus calamus*), which was cultivated for its scent. Dried and crushed, the leaves release a cinnamon-like perfume. The roots were also dried and powdered and used as a type of talcum powder for perfuming the body and linen, and the leaves were macerated to prepare aromatic oils and vinegar.

Imperial Rome emulated the Greeks' love of scented plants. Many of the flower essences they favored were so popular that their manufacture became a flourishing industry, for example, *rhodium,* made from roses; *melinum* made from quince blossoms; and *narcissinum,* from the narcissus. *Susinon,* made from saffron, calamus and oil of lilies, was widely used in the public baths and barber shops for massaging into the skin after shaving.

Although the Romans were less aware of eastern aromatic spices and gums, they did have a passion for perfumed flowers. Every spring, a festival was held in honor of the goddess Flora and the first budding May branches were twined into garlands and hung on house doors as a sign that the season of fruitfulness had arrived. The saffron crocus (*Crocus sativus*) was much favored by the Romans. The flowers were strewn in ceremonial halls and on banqueting tables and the powdered stigmata were used in a fragrant dye to tint hair a bright blonde. Saffron was renowned for its medicinal and culinary properties and was also used to flavor and tint flour. This gave the pastry a rich, golden color, which was thought to be very lucky. In later times it came to be highly prized as a fabric dye.

The Romans also loved roses and used them lavishly. Rosewater went into their baths, flowed from palace fountains and was liberally splashed about the amphitheaters to disguise the gruesome stench of blood. At banquets and festive occasions, roses were scattered everywhere. Cascades of the petals were showered from the ceiling to such an extent that some guests were apparently suffocated by the avalanche of falling petals. Roman wine was flavored with roses, as was food and medicine.

Cato wrote that violets and the Imperial Crocus (*Crocus imperati*) were widely grown in Roman times, their flowers being woven into chaplets or wreaths for daily wear by Roman women. Roman men also used aromatic plants, particularly bay leaves, which they placed in their baths to relieve battle-weary limbs. The Romans made great use of their baths and the *thermae* were an important part of the city's social life. In the early days, the Romans anointed their bodies after bathing only with pure olive oil. However, in tandem with the growth of excess in their culture, perfumed oils were gradually preferred. Julius Caesar criticized

▸ A garden gazebo framed by magnolias, pink dogwood and violas.

*Crocus* sp.

this use of perfumes for personal pleasure as effeminate and decadent and forbade their sale. He was reported as saying to a particularly highly scented fop, 'I'd rather you stank of garlic'! However, by the time of Nero, perfume consumption had reached an all-time high in Roman society. Poppaea, the seductive and corrupt wife of Nero who, according to Tacitus 'had every asset except goodness', bathed daily in scented asses' milk, keeping a herd of 500 she-asses for this purpose. Poppaea's fashion foibles kept upper-crust society in a dither and Petronius dourly observed:

*Wives are out of fashion. Mistresses are in.*
*Rose leaves are dated,*
*Now cinnamon's the thing.*

In Nero's state apartments each room was carpeted with red rose petals every day. Nero also had silver pipes built and hidden in the walls. At his signal, the servants would pull levers that shifted moveable ivory plates on the outside of the pipes and every guest would be showered with scented water. At Poppaea's funeral, Nero was said to have ordered more fragrant woods to be burnt than would be produced by Arabia in one year.

A Very Good Perfume to Burn

*Take two ounces of the Powder of Juniper Wood, one ounce of Benjamin, once ounce of Storax, six drops of oil of lemons, as much oil of Cloves, ten grains of Musk, six of Civet, mould them up with a little gum-Dragon steeped in Rosewater, make them in little cakes and dry them between Rose-leaves, your Juniper Wood must be well dried, beaten and searced.*

W. M. (Cook to Queen Henrietta Maria), *The Queen's Closet Opened,* 1655

One of the most popular of all perfumed plants was named after a famous Roman family, several centuries after the fall of that Empire. When Mercutio Frangipani voyaged with Christopher Columbus to the New World in 1492, he made a discovery of his own in the West Indies—the white *Plumeria alba* and the rosy-pink *Plumeria rubra,* which henceforth bore his name. The dried flowers were combined with the thick juice of *Cascarilla gratissima* and made into a pomade, which Frangipani advocated be rubbed into gloves. This fashion for scented gloves was to continue for centuries. It is recorded that Queen Elizabeth I was presented with several pairs by the Earl of Oxford on his return from Italy and she was extremely pleased with them.

The New World was to prove a treasure trove of scented flowers and shrubs: exquisite magnolias came from Louisiana, the cinnamon-scented Sweet Fern (*Comptonia aspleniflora*) from North America and the West Indian Bay (*Laurus sassafras*), which became an important ingredient in Bay Rum, a famous men's hair tonic.

Another fragrance used around the world is oil of wintergreen, obtained from the Ground Holly

*While wormwood hath seed, get a bundle or twain,*
*to save against March, to make flea to refrain:*
*Where chamber is swept, and wormwood is strown,*
*No flea, for his life, dare abide to be known.*

Thomas Tusser, *Five Hundred Points of Good Husbandrie,* 1573

❖ ❖ ❖

(*Gaultheria procumbens*), which comes from North Carolina.

With the fall of the Roman Empire, the power and the passion for perfumed flowers and their byproducts once more moved to the eastern empires. Constantinople was a splendid, scented city; its palaces were full of fragrances and its ports were the centers of the perfume trade. The Arabs became perfume suppliers to the world, following to the letter Mohammed's philosophy that the three most important objects of desire in the world were 'women, perfume and prayer'. It was the great Arab doctor and philosopher, Avicenna, who made the most important discovery in the history of cultivating plants for their perfume—alcohol. By inventing the process of distillation, he set the path for discovering modern perfumes. His experiments were conducted on the Arabs' favorite flower, the rose, and the resulting product, rosewater, is still a worldwide favorite.

Rosewater was introduced to Europe as a result of the Crusades. Christian knights became fascinated with eastern ways, particularly the idea of the harem. On their return, the gullible knights hopefully proffered their womenfolk the flower waters that, they had been assured, were favored by the sensuous women they had seen on their travels. The Damask Rose (*Rosa damascena*) took its name from the city of Damascus, the ancient capital of Syria. Interestingly, 'Syria' is thought to have derived from *suri*, meaning 'land of roses'. The Damask Rose was introduced to Europe by returning Crusaders and began to challenge the popularity of the classical red rose, *Rosa gallica,* one of the oldest plants known to humanity.

The knights also brought with them the seeds of Love-in-the-Mist (*Nigella sativa*) and various scented lilies and irises, notably the Madonna Lily

*Acorus gramineus* 'Sweet Flag'

(*Lilium candidum*). The silent 'war' between the lilies and the roses started around this time, with both being regarded as 'Queen of the Flowers'. Roses and lilies were also imbued with religious significance and appeared often as motifs in medieval art.

In early European homes, scented flowers and herbs were primarily used for strewing on the floor to reduce the unpleasant stale odors that often resulted from the damp floors and very small windows. In his *Five Hundred Points of Good Husbandrie* (1573), Thomas Tusser expressed the view that basil, balm, chamomile and costmary, lavender, hyssop, sage and thyme were all 'most excellent' for use as strewing herbs. Freshly gathered scented rushes (*Acorus calamus*) were also

*Let's go to that house, for the linen looks white and smells of lavender, and I long to be in a pair of sheets that smell so . . .*

Izak Walton, *The Angler,* 1653

popular, but due to their high price, they were only to be found in the homes of the gentry. In Pegge's *Curalia,* it was recorded that the rather profligate Thomas A'Becket ordered his hall 'to be strewed every day, in the Spring, with fresh May blossom, and in summer with Sweet-Scented Rushes . . . that such knights as the benches could not contain might sit on the floor without dirtying their clothes.'

The rich scent of damask roses was used in a variety of ways to clear the fetid atmosphere. According to *Les secrets de Maistre Alexis Piedmont*, a very early French treatise on perfumes, a fine 'Damask scent' to fumigate a room could be made by combining five grains each of musk and ambergris; two grains of civet; of fine sugar, four grains; benzoin one grain; and storax, calamus and aloes-wood, three grains each. This mixture was then placed in a perfume pan and covered with 'Damask Rose-Water . . . two fingers high' and burnt, filling the room with an aroma of incense-like perfume.

Cloves, the dried fruits of *Eugenia caryophyllus*, were similarly burned to scent rooms. This custom was recorded centuries before the birth of Christ, when it was a favorite fragrance in the Chinese Han Court. Envoys to the Palace were ordered to place cloves in their mouths before addressing the court, so as not to taint the air. In addition to their spicy fragrance, cloves provided a therapeutic benefit, having strong antibacterial properties. Through ancient and medieval times, cloves were imported from the Middle East to Europe and used to pickle and preserve meats. In the seventeenth century, sponges soaked in clove-scented oil were held beneath the noses of victims of the plague. Doctors tending the victims attempted to keep the disease at bay by breathing through an odd type of 'mask' that had a beak filled with cloves.

Sweet-smelling woods, such as pine, cypress, juniper and fir were burnt to provide both warmth and fragrance. Dried and then burnt over a low flame, elecampane roots produced a violet-like perfume and the aromatic seeds of the angelica plant (*Angelica archangelica*) were also recommended as an all-purpose fumigant by Stevenson in his *Calendar for Gardening* (1661):

*. . . be sure every morning to perfume the house*
*with angelica seeds, burnt in a fire-pan or chafing*
*dish upon coales . . .*

The eminent Dr Turner made references in his *Herbal* to the rather alarming incidence of 'serpents' that lurked in dark corners about the home. We may suppose he was meaning frogs and toads, however, the snake-phobic should remember that burning aromatic southernwood will drive away any noxious creature.

The great period for perfumed flowers and herbs in England was during the reign of Elizabeth I, the 'Virgin Queen'. It was also at this time that more formal garden styles, which capitalized on the fragrances of different plants, began to emerge. Bowers of scented vines, such as honeysuckle and jasmine, were especially popular and Ben Jonson immortalized this quintessential feature of an English cottage garden in his *Vision of Delight*:

*So the blue bindweed doth itself enfold*
*with honeysuckle, and both these entwine*
*themselves with briony and jessamine,*
*To cast a kind and odoriferous shade . . .*

The herbalist John Parkinson, writing at about the same time, recommended that 'the jasmine, white and yellow, the double honeysuckle, the Ladies' Bower, white and red and purple, single and double' be planted about an arbor or bower. Thomas Hill, in *The Profittable Arte of Gardenynge* (1563) preferred '. . . that sweet tree of flower named Jacemine, Rosemary or Pomegranate seed unless you would rather deck your arbors comelier with vines'. The word 'arbor' came from *herba*, meaning a garden of fragrant plants '. . . so pleasant and so dulce, the pestilent ayres with

### A Sweet and Delicate Pomander

*Take two ounces of Labdanum, of Benjamin and Storax one ounce: musk, six grains; civet, six grains; Amber-grease, six grains; of Calamus Aromaticus and Lignum Aloes, of each the weight of a groat; beat all these in a hot mortar and with a hot pestall until they come to a paste; then wet your hand with Rosewater and rowle up the Paste suddenly.*

Sir Hugh Platt, *Delightes for Ladies*, 1594

❖ ❖ ❖

*Rosa* 'Pascali' (Hybrid Tea)

flavors to repulse'. Clipped, stylish hedges of rosemary and lavender surrounding low beds set out in 'knotte' or maze designs were also popular, as were lawns of scented herbs, such as chamomile and thyme.

Queen Elizabeth I was reportedly fascinated by different scents. Said to possess a very sensitive nose, she was unable to bear unpleasant smells and spent very lavishly on flowers and perfumed waters, paying her stillers, John Kraunkcell and his wife, the then extremely extravagant sum of 40 pounds per year for their waters '. . . most swete and delicate . . . of violette and gilloflowre.' The latter was the Elizabethan name for clove-scented pinks (*Dianthus caryophyllus*), introduced to England by knights returning from the Crusades. It was a particular favorite in England at this time, and was used for flavoring wine as well as making perfume. The petals of Pinks were the 'sops-in-wine' often referred to by both Shakespeare and Chaucer.

Fragrant herbs and flowers were also used to impart their scent to the Queen's linen and repel moths from her furs. Her clothes were stored in chests made from ornately carved aromatic wood, such as cedar. Jasmine buds, rose petals, bay leaves, dried orris root, cloves and dried imported frangipani petals were all popular with Elizabethan housekeepers, as were mint, thyme and rosemary.

Lavender, long prized by the Romans for its clean, refreshing scent, was distilled in many sixteenth-century English country houses to produce lavender water. It was so popular for perfuming clothes that a laundress was quite likely to be nicknamed 'the lavender'.

Pots of lavender and mint were set at every window to perfume the often fetid air of early England. 'Buy my sweet lavender, two bunches a penny', was one of the street cries of flower sellers in early London. The dried leaves and flowers were used to fill pillows and mattresses, along with dried

*Take a gallon of faire water, one handfull of lavender flowers, a few cloves and some Orace powder and foure ounces of Benjamin; distill the water in an ordinary leaden still. You may distill a second water by a new infusion of water upon the leavesl a little of this will sweeten a bason of faire water for your table.*

Sir Hugh Platt, *Delightes for Ladies,* 1594

agrimony and woodruff and hops, the latter having a mildly sedative effect. Rosemary was also thought to encourage sound sleep and bags of the dried leaves and seeds were hung over beds. Widely enjoyed for its cosmetic and culinary benefits, rosemary appeared in every English garden. Sir Thomas Moore wrote of it:

*As for rosemarine, I let it run over my garden walls, not only because the bees love it, but because it is the herb sacred to remembrance, to love and to friendship.*

Tudor times were rich with fragrance, emanating from many scented waters, powders, potpourri, pomanders and gilt cassolettes (intricately carved and perforated boxes used for burning incense). In addition to the traditional pomander (a dried orange or apple stuck all over with cloves and/or spices thought to offer protection against disease and vermin) many more elaborate ones were fashioned from gold and precious jewels. One of Queen Elizabeth's courtiers presented her with '. . . a fayre gyrdle of pomanders', that is, a cluster of tiny pomanders, each with six orange-shaped segments made of filigree silver, which could each be filled with different potpourri or scented oils.

Potpourri preparation became as popular a hobby as it is today. Ladies of leisure set aside whole rooms to experimenting with different blends, inviting guests to participate in potpourri-making bees. In his *Delightes for Ladies* (1594), Sir Hugh Platt had this advice for his gentle readers: '. . . hang your pot in an open chimney or near a continual fire so that the petals will keep exceedingly faire in colour and be most delicate in scent'.

### To Make Oyle of Roses

*Take of oyle eighteen ounces, the buds of Roses (the white ends of them cut away) three ounces, lay the Roses abroad in the shadow four and twenty houres, then put them in a glass to the oyle, and stop the glasse close; and set it in the sunne at least forty days.*

John Partridge, *The Treasurie of Hidden Secrets and Commodious Conceits,* 1586

Exquisite casting bottles of chastened silver were much favored by court gallants and their ladies, who would sprinkle the fragrant waters within liberally over the floor of any chamber they entered. A recipe from the dashing Sir Hugh Platt for a 'swete water' contained rosewater, three drams of spikenard oil, one dram each of oil of thyme, lemon and cloves, plus a grain of civet. In Marlowe's *Faustus,* the good doctor wrinkles his nose and says: 'Fye, what a smell is here, I'll not speak another word for a king's ransome, unless the ground is perfumed'. Similarly, pretty porcelain dishes of rosewater were placed in dressing rooms and next to guests at the banquet table, so their hands and face might be readily refreshed and perfumed. Shakespeare refers to the use of these scented preparations as a tonic and hangover cure when he writes of the remedy planned for the drunken Christopher Sly in *The Taming of the Shrew*:

*Carry him gently to my fairest chamber,*
*Balm his foul head in warm distilled waters,*
*And burn sweet wood to make the lodgings sweet:*
*Let one attend him with a silver basin,*
*Full of rosewater.*

Aromatic snuffs, comprising dried and crushed chamomile leaves, peppermint, woodruff and pellitory were used by all classes of Elizabethan society, though snuff-taking did not reach its height until the Stuart era. The first cigars also came into use about that time, with John Gerard providing explicit instructions for their use: 'The dry leaves are used to be taken in a pipe and set on fire and sucked into the stomach and sucked forth again through the nostrils.' Interestingly, the word 'cigar' is believed to have come from 'cigarral', the Spanish name for 'little garden', due to the Spanish custom of growing a *Nicotiana tabacum* in every garden.

A Sweet-smelling Perfume

*Take a pound of fresh-gathered Orange flowers, of common Roses, Lavender seeds and Musk Roses, each half a pound; of Sweet Marjoram Leaves and Clove-July flowers picked, each quarter of a pound; of Thyme, three ounces; of Myrtle leaves and Melilot Stalks stripped of their leaves, each two ounces; of Rosemary Leaves, and Cloves bruised, each an ounce; of Bay Leaves, half an ounce.*

*Let these ingredients be mixed in a large pan covered with parchment and be exposed to the heat of the sun during the whole summer for the first month, stirring them every other day with a stick, and taking them within doors in rainy weather. Towards the end of the season they will afford an excellent composition for perfume; which may be rendered yet more fragrant, by adding a little scented Cypress-powder, mixed with coarse Violet Powder.*

*The Toilet of Flora*, 1775

Certain flowers, herbs and seeds were used to polish furniture, providing a glossy sheen and long-lasting fragrance. In *The Merry Wives of Windsor*, Shakespeare refers to the use of balm (*Melissa officinalis*) in this way:

*The several chairs of order look you scour*
*With juice of balm, and every precious flower:*
*Each fair instalment, coat and several crest,*
*with loyal blazon, evermore be bles't.*

The fastidious Queen Elizabeth I favored the delicate meadowsweet above all other strewing herbs. Of this dainty herb, Gerard said:

*The leaves and flowers far excell all other strewing herbs to deck up houses, to strew in chambers, halls and banqueting houses in summer time, for the smell thereof makes the heart merry and joyful, and delighteth the senses.*

*Melissa* sp.

The practice of strewing fragrant herbs and grasses continued in England until the late eighteenth century, and still has a symbolic importance at certain ceremonies. For example, at the coronation of George IV, we are told that a certain Miss Fellowes, sister of the Secretary to the Lord Great Chamberlain, was appointed as the Herb Strewer for that occasion. Dressed in white muslin and 'flower ornaments', she was charged with scattering scented blossoms in front of the royal procession.

Strewing eventually fell from favor because the stale herbs and flowers provided an ideal breeding place for vermin, necessitating the regular practice of burning the rather foul-smelling Fleabane (*Erigeron acris*) to deter the pests. This awareness of the hazards of strewing arrived too late for most Londoners—sadly the practice of carpeting houses

A living bouquet of colors and aromas; the small-flowered Kurume azalea and *Primula malacoides* 'Gilhams White' provide a solid background for the many shades of these mixed pansies.

and taverns with 'swete herbes' was a key factor in bringing about the Black Death and, later, the Great Fire of London.

Madame de Pompadour, mistress to Louis XV, lavished perfume upon herself and filled her apartments with scented flowers on an enormous scale. She disliked spicy perfumes and heavy civet-based scents, preferring to set a fashion for lighter, floral fragrances. Her favorite flower was the hyacinth, especially the double variety, and contemporary historians wrote that she ordered masses of them to be cultivated indoors during winter and spring when few other flowers were available. The bulbs were grown in exquisite glass or porcelain containers, shaped so that the bulb's base just touched water; no soil was used. The bulbs were forced in darkness and burst into flower when they were taken to Madame's warm boudoir where they retained their perfume for many weeks. Madame de Pompadour succeeded not only in filling her rooms with fragrance, but in making hyacinths so popular that their price skyrocketed.

This change in fashion survived the upheaval of the French revolution. Like Madame de Pompadour, Napoleon's wife Josephine was also fond of keeping pots of scented flowers in her rooms, particularly hyacinths and mignonette, or 'Little Darling'. During his Egyptian campaign, Napoleon arranged to send some mignonette seed home to her and it rapidly became popular in French society. Soon it was very fashionable to grow mignonette in pots on balconies to mask the putrid odors that rose from the busy, crowded streets. Josephine also had a passion for violets, that elusive fragrance so poignantly described by Shakespeare as being:

> *. . . forward, not permanent; sweet, not lasting.*
> *The perfume and suppliance of a minute.*

Josephine cultivated many different varieties of violets at her famous garden, Malmaison. When she died, Napoleon asked for her grave to be planted all over with violets. Shortly before his exile, he plucked a few flowers from her grave which were found in a locket about his neck when he died.

Eau de Cologne was also a popular scent in Josephine's time. Early in the eighteenth century, Paul Feminis, an Italian living in the German city of Cologne, prepared a toilet water based on citrus oils, neroli, lemon, bergamot and lavender. Nearly a century later, one of his descendants added rosemary to the formula and created Eau de Cologne. This was Napoleon's favorite perfume and he used it extravagantly throughout his life. Even when on campaign, he was known to spread it lavishly over his neck and shoulders each time he bathed. It is said the rosemary in the scent appealed most to Napoleon, because it reminded him of his boyhood in Corsica and the rosemary that grew, and continues to grow, wild in the maquis. Eau de Cologne was an immediate success and helped to establish the trend towards soft, fresh, natural, plant-based fragrances, which has lasted to today.

The natural elegance of a garden archway.

PLANTING A

# Scented Garden

A garden of aromatic herbs and flowers is uplifting to the spirit as well as to the nose. Apart from the constraints of your own imagination, there is no limit to the placement of perfumed plants—from window-boxes and rustic balcony tubs to a paved courtyard with raised beds; from lovely old beds of the traditional favorites, such as lilac and daphne, to more formal 'knotte' gardens of summer-warmed herbs.

Let the spirit of romance enter your planting plans and make your scented garden a place where you can retreat from the daily hurly-burly of the world. Wreathe forbidding brick walls or boring fences with festoons of jasmine and old-fashioned Bourbon roses. Set trellised archways or screens near your house and encourage scented, blossomy climbers to clamber up to your windows.

◂ With creative variations on color and scent, decorative path borders will enhance your garden amble.

If the budget permits, consider hiring an excavation team for a day to create a glamorous water garden in a sunken or damp area of the garden. Heap up the excess soil and rubble into soft mounds and create a miniature 'landscape' planted all over with scented carpeting herbs and, perhaps, a copse of exquisite, perfumed white blossom. The smallest pond will help moisten the air and provides an ideal spot for displaying scented rushes or irises and even exotic waterlilies, if the climate permits. *Nymphaea odorata* 'Alba' needs waist-high water in order to flower, but many other varieties, notably *Nymphaea pygmea* 'Helvola', which has sweetly scented flowers of palest yellow, will bloom in a bowl of water only 12–18in/30–45cm deep.

If you have no room for a watergarden, water may still be introduced using large and elegant Chinese ceramic urns or wooden half-barrels lined with sealant to prevent leakage. And to anticipate the inevitable concern about water attracting

Strewing Herbes of all Sortes

*Bassell, fine and busht, sowe in May*
*Bawlme, set in Marche*
*Camamel, Costemary Cowsleps and paggles Daisies of all sorts*
*Sweet fennel Germander Hop, set in Februarie*
*Lavender lavender spike Lavender cotten*
*Marjoram, knotted,*
*sow or set, at the spring*
*Mawdelin Peny ryall Roses of all sorts,*
*in January and September Red myntes sage Tansey*
*Violets Winter savery*

Thomas Tusser, *Five Hundred Pointes of Good Husbandrie*, 1573

mosquitoes, add a couple of fish to your pond or urn—they love eating mosquito larvae!

When planning a scented garden it is important to ensure that you have a seasonal balance of trees and shrubs, annuals and perennials, so that something is always flowering throughout the year. Many scented plants mentioned in this book are winter-blooming and they add color and fragrance to a gloomy day. Also certain varieties of the individual species, notably the scented liliums, bloom at different times of the year, making them an excellent choice for bedding.

Also take great note of the color choices available to you. Because they are usually night-pollinated, many scented plants are white or cream-flowered, which does not create a problem if you are planning an all-white garden, or even a pale-pink and gray garden. However, if you are seeking color as well as scent, you will need to remember to plant brighter varieties as well as the paler, scented ones. A clever planting choice for a pergola or summer-house would be a scented Banksia Rose (spring flowering) and a yellow Cape Jasmine (summer flowering), interplanted with a golden bougainvillea which, although not scented, will flower right through to the end of autumn in a warm, sheltered position. Even though bougainvillea is not scented, the dried flower bracts provide much needed color and texture in potpourri or other dried flower arrangements.

*But flowers distilled, though they with winter meet,*
*Lose but their show, their substance still lives sweet.*

William Shakespeare

## Paths and Surfaces

Paths are often underestimated and regarded merely as a means of getting from one part of the garden to another. But a path can be much more than merely functional. You can create a shady, scented walkway that does not necessarily go anywhere—except possibly to a nook beneath a fragrant lobelia, to a secluded pond or to a favorite stone birdbath, placed in a quiet corner to attract feathered friends.

Wide gravel paths may be canopied with perfumed climbing plants and edged with low-growing plants that release scent as they spill on to the path and are brushed against. Weathered flagstones are charming set about a timber cottage, though river pebbles or terracotta tiles may be a smarter and more appropriate choice for a modern home. Paths should be planned to complement the vintage of a house as well as its dimensions and the way in which the resident family uses it. Old bricks, placed on their sides, make a distinctive path. Two-inch square gaps should be left randomly and planted with aromatic herbs of a

*Endeavour to make the principle Entrance into your Garden, out of the best Room in your House, your walks being places of divertisement after a sedentary repast. The Aromatick Odours they yield, pleasant refreshments after a gross diet, and such innocent Exercises the best digestive to weak Stomacks. And let your principle Walk extend itself as far as you can . . . adorned with the choicest Plants for Beauty and Scent, and that there may be a succession of them throughout the year, not without flower pots, which grace the best of gardens.*

John Worlidge (1677)

This cobblestone path is surrounded with scent: heather lines each side of the path, while colorful lupins provide additional height. At the end, white daisy bushes flank a classic thyme ball arrangement.

carpeting habit. Stone pavers, shingle and gravel also look most attractive. One trick to remember is to allow the path to meander. As well as being far more interesting than a straight path, a curved path will produce a greater impression of space in the garden.

Do not be too conservative when planning a pathway. A suburban plot can be transformed with a little thought. A friend in England, who dearly loved *The Wizard of Oz* as a child, has created a whimsical path of yellow pebbles through a 'secret' corner of her garden. The path is fringed with buttercups and wild strawberry runners, eliciting exclamations of delight from children and adults alike. The use of such coloured pebbles was fashionable during Tudor times, and can still be enjoyed for the very satisfying crunching noise they make underfoot.

Keeping paths narrow will create a more intimate, romantic feeling, but be generous with your planting by paths. Make flower beds as wide as possible and do not be mean with borders. And, if you have the ubiquitous sun-parched front lawn found in far too many suburban homes, why not rethink that space? Pretty shrubberies linked by pathways are far more attractive, give more privacy and, if well mulched, are a lot easier to maintain than a lawn.

In 1625 Francis Bacon described wild thyme and water mint as '. . . perfuming the air most delightfully, being trodden upon and crushed.' Why not heed his advice and plant these, and other different aromatic herbs by and in your path. Set plantlets about 3ft/1m from each other, for they will soon spread out to cover an area two foot square, or more. Planting different species or types of plants every yard or so offers new interest to the passer-by strolling along the path. Do not overcrowd your herbs when planting as they will need plenty of room to spread out.

Among the best choices are the prostrate thymes, especially *Thymus serpyllum,* which produces a mat of dark green foliage dotted with masses of tiny little red flowers, or Mother of Thyme, *Thymus praecox,* which has delicate pinkish-purple flowers. Pink Chintz, with salmon-colored flowers, is also pretty.

If, like me, you find yourself attracted to collections of herbs and flowers, you might like to think about massing clusters of the different types of pinks, mints (for example, apple mint, spearmint, peppermint) or cat mints along the sides of your path. The perfumes that blend and rise to greet you—each different yet complementary—make for a heady mixture. Although most herbs enjoy dry conditions, mints like plenty of moisture and benefit from a little rotted manure tucked about the plants' roots.

Complement mint with the non-flowering dwarf chamomile, *Anthemis nobilis* 'Treneague', which releases a clean, apple-like scent when it is walked upon. In Spain, it is still known as Manzinella, which means 'Little Apple'.

*Anthemis* sp.

Other dwarfish plants to set alongside or just back from the path include the small irises, rosemary and perennial violas, and, in the sunniest places, one or two of the fairy miniature roses. It is also hard to surpass lavender for rich scent and color. But do not limit yourself to just one or two lavender bushes—why not create a colorful display by planting a double, or even triple, hedging of white, pink and traditional blue lavender, with the high bushes at the back and dwarf varieties in front. This can be quite spectacular, especially if you choose varieties that bloom at different times so the flowers open in succession and attract bees through summer.

## A Chamomile Lawn

The lawn should also be part of a scented garden. An aromatic carpeting herb, such as chamomile, thyme or mint, makes a refreshing change from couch or kikuyu. A perfumed lawn really comes into its own in a small city garden where space is at a premium and noise and traffic smells are prevalent. It is also an asset in little-used nooks in the garden, where an idle browser will get a pleasant surprise if they step on the scented carpet.

Chamomile was used in Tudor times to make an aromatic lawn and as grass lawns were not common in this era it is more than likely that Drake played his famous game of bowls upon one. The first grass lawns, as we know them, were not planted widely until the eighteenth century. Buckingham Palace sports the remnants of a chamomile lawn, although over-zealous mowing is thought to have deterred growth and flowering.

*And let the ground whereas her foot shall tread*
*For feare the stones her tender foot should wrong,*
*Be strewed with fragrant flowers all along,*
*And diapered like the discoloured mead.*

Edmund Spenser, seventeenth century

To make an aromatic chamomile lawn, use the variety *Anthemis nobilis* 'Treneague', which has a fairly rapid spreading habit. First, mark out the area for your lawn and scatter seed sparingly—a handful produces many hundreds of plants. Rake the seed into the ground and water gently. Keep the soil moist as the plants come up and thin them out when they are established to about 4in/10cm apart, moving excess seedlings to other spots where the seed has not taken as well. If you manage to keep the area weed-free during the first year, weeding will be greatly reduced in seasons to come, for as the chamomile spreads out in a thick 'mat', it suppresses weed growth.

By spring, your chamomile lawn should be ready to be walked upon. To trim, most experts suggest clipping with garden shears rather than mowing, at least for the first year or so, as mowing can damage the plants. If you do mow, then only do so once every three months. This will encourage the plantlets to thicken up and, by the second year, you may enjoy the fruity perfume emitted from your chamomile lawn.

Thyme is another good choice for carpeting a steep or inaccessible patch in the garden. While a traditional lawn might be hard, or impossible, to mow, a thyme lawn would only require spasmodic trimming. Wild Thyme (*Thymus serpyllum*) flows over parched sandy soil and rocky outcrops in a most satisfying and vigorous way, turning undesirable landscapes into scented highlights.

*Mentha requienii*

While thyme is best suited to a hot or dry aspect, a mint 'lawn' will do well in semi-shaded, moist areas in a garden. *Mentha pulegium*, or Pennyroyal, is the variety often used for lawn-making. It has dainty pale mauve flowers and is extremely close-growing, forming a thick, springy fragrant mat. Mints, however, will spread most enthusiastically, so if you only want them in one part of the garden, it is best to contain your mint 'lawn' with bricks or tile edging.

## Scented Seats

Fragrant benches, set with creeping thyme or chamomile that released their scent when sat upon, were popular features in Elizabethan gardens. It is a reasonably simple matter to construct a fragrant seat and it will allow you to contemplate the beauty of your garden in a haze of fragrance. A simple bench may be built against a brick wall or the side of the house, by building a 'trough' from either bricks or railway sleepers and filling it with soil before planting out with the scented ground cover of your choice. Once again, creeping chamomile, *Anthemis nobilis* 'Treneague', is one of the prettiest choices for such a seat. At her famous herb garden at Sissinghurst, Vita Sackville-West had a chamomile seat where she was wont to sit, eyes closed, inhaling the many fragrances that filled her garden.

## Raised Beds

Some of the sweetest scented plants are also the smallest, such as tiny violets, scented cyclamens, dwarf savory or primroses. People with bad backs, older or disabled gardeners will benefit most from having such plants brought to a manageable level via raised beds. Such raised gardens are also delightful for people who do not like to miss out on anything—everyone can delight in the small fragrant plants without having to get down on hands and knees.

There are several ways to create a raised garden. Avid 'garage-sale' attendees should keep their eyes open for old claw-footed baths or, those rare treasures, stone watering-troughs and old-fashioned enameled or porcelain sinks. There are also many attractive terracotta troughs, in varying sizes, available at your specialist gardening shop or nursery. The great advantage of 'trough-gardening' is that you can work the soil differently, using a very acid mixture in one, an alkaline mixture in another, and so on, which will provide a controlled environment for sweet-smelling plants that may not tolerate ordinary garden soil. For example, the prostrate *Daphne cneorum*, which normally creeps along the edges of borders, will grow well in a lime-enriched trough and also be far more visible.

Building raised beds from bricks and including pockets of soil within the walls will provide a lot more planting space in a small city garden or courtyard. Lower pockets can be planted, or 'padded', with cushiony mints or thyme and used as seats while the upper pockets can take taller plants and shrubs that like well-drained soil, so providing shade and privacy. Always mulch raised garden beds or troughs thoroughly as they will tend to dry out quickly, far more rapidly than other areas in the garden.

Use raised beds to create a 'sea' of different scents; plant lavender with aromatic bergamot, pelargoniums, orange thyme and rosemary for a spicy effect. Or, for a sweetly-scented bed, English wallflowers complement daphne, night-scented stock and violets equally well.

## A Scented Bower

Covered with a mass of climbing roses or flowering vines, it is not surprising that bowers have long been a popular spot for sweethearts to meet! They are also a wonderful retreat and place to share the garden with family and friends. Imagine lunch, afternoon tea or a candlelit dinner in a bower set in the middle of a fragrant herb garden. When not in use, a bower is an ideal place to house a collection of shade-loving plants.

A bower may be arched or square, made of rustic twigs, carefully painted wood or even sturdy, maintenance-free plastic. The floor could be made of gravel, sand, or be paved with flagstones. Place a chair or bench of rough-hewn wood, or an

A secluded garden seat is sheltered by an arch of rambling 'Dorothy Perkins' roses, and overlooks a chamomile lawn fringed with pinks and violas.

artfully designed wrought-iron couch in your bower. Simple stone seats could also be used and 'upholstered' with judicious planting of creeping chamomile or thyme.

If possible, build your bower where you have a view of the whole garden or in a secluded corner near a well-placed shady tree. Ideally the aspect should face west so you can enjoy watching the sun set at the end of your day's work in the garden.

In his *Records of the Plants of Southern China*, the third century historian Chi Han wrote that Canton in the jasmine season was ' . . . like snow at night, fragrant everywhere.' Jasmine flowers were cultivated widely in ancient China and used primarily to make perfumes and tea. As befits a plant with such exotic connotations, the dainty primrose-like flowers of *Jasminum officinale* are a fitting choice for a romantic, scented bower. The Elizabethan poets mentioned it as being used for trailing over arbors. Jasmine grows vigorously and makes plenty of bushy growth. Tie in the stems to the trellis or training wires over the bower as it grows. To protect the plant's roots, set rosemary or lavender at its base and, to increase the heady fragrance still further, train a tobacco plant, *Nicotiana alata grandiflora*, about the arbor. Brazilian girls were said to wear the pink flowers of tobacco plant in their hair as an aphrodisiac.

Every romantic bower should have at least one of the honeysuckles growing upon it. Due to its habit of tightly clasping its support, honeysuckle has been known as 'love-bind' or 'hold-me-tight' in country circles for many years. Try planting two at the back of a bower so they may climb together. *Lonicera periclymenum* blooms in high summer. Perhaps the loveliest of all is *L. japonica*, with its sweetly-scented creamy yellow flowers.

In his *Herball*, John Gerard called clematis 'Virgin's Bower', and wrote that it was once a popular choice for protecting young girls from the rude gazes of passers-by. *Clematis vitalba* is a lovely choice for a scented bower, with its clusters of greeny-white flowers and masses of autumnal feathery gray seed pods. The latter give the climber a soft and misty appearance at the end of summer, hence the derivation of its nickname 'Old Man's Beard'. *Clematis flammula*, or 'Fragrant Bower', is also gorgeous, with large clusters of creamy white flowers that have a sweet and delicious scent.

The bright gold leaves of Golden Hop, which takes its name from the old Anglo-Saxon word *hoppan*, meaning 'to climb', look very striking next to the clematis' soft flowers.

Climbing scented roses, of course, need little introduction. For a scented bower or arbor, or for training along a wooden fence or a brick wall, the rich sweet scent of the pink rambler rose 'Albertine' is hard to surpass. Again, using a perfumed climber as a partner can create a spectacular display, as opposed to one that is just very attractive. The Giant Honeysuckle, *Lonicera hildebrandiana*, has richly sweet, creamy flowers, which create the perfect foil for a pink rose.

## Window Boxes

As long ago as 1594, Sir Hugh Platt wrote: ' . . . in every window you may make square frames . . . of boards, well pitched within. Fill with . . . rich earth and plant therein such flowers and herbs as you like best.' His idea remains as fresh today and, where space is limited, window box gardening is a terrific idea for growing aromatic plants, either for the kitchen or for scenting the home.

A wide range of scented plants may be successfully cultivated in window boxes and they will have a cheering effect on the house, both inside and out. Thyme, marjoram and winter savory, the tiny dwarf lavenders and golden sage are all pretty choices for window boxes and, being evergreen, will remain looking good all year round.

If you want even more color, set one or two plants of bright red or pink geraniums in the window box as well. The traditional approach is to feature flowers that signal the seasons, such as stocks, spring-flowering bulbs, hyacinths and primroses. Try to select plants that harmonize well with their surroundings and think about the color combinations you are creating to best ensure a pleasing result. For example, warm yellow wallflowers with dainty blue forget-me-nots; rich blue violets with white or yellow hyacinths; or a mass of primroses and red wallflowers will all provide a colorful display and sweet scents at your window. Miniature roses, such as 'Yellow Doll' (only 12in/30cm high with palest cream blooms), are especially charming in a window box.

*I think nothing can be more delightful than to throw open your window and to inhale a refreshing odour from growing flowers when they are swept over by a balmy breeze . . .*

Mrs Earle, *Potpourri from a Surrey Garden*, 1905

Different types of scented pelargoniums are lovely when used around a verandah in window boxes, where they are likely to be leaned against or brushed past. The rose-scented variety has a pronounced bouquet as does the peppermint-scented one; both are quite vigorous and positively thrive in hot dry conditions. For a window box perched outside the kitchen, what could be more appropriate than a selection of freshly scented herbs? Basil, angelica and mint will form a display pleasing to both eye and nose.

Victorian window boxes made of metal with ornately carved patternwork on the front are much sought after by home renovators at present. If you are unable to find these, though, many of the newer plastic ones or wooden ones are quite satisfactory and come in a variety of colors, styles and lengths. Use fresh, rich loam to fill the boxes and add a few generous dollops of manure or blood and bone to give plants the necessary impetus to flower. Spring is the best time to plant window boxes. A good tip is to regularly pinch back new growth so that growth is bushy rather than straggly.

## Hanging Baskets

In Elizabethan times, housewives would tie jars of rosemary, which grows very well indoors, to the outside of the chimney breast in the kitchen during summertime to keep the air cool. This practice of hanging baskets of aromatic herbs or flowers indoors to freshen a room is still used today.

Whether it is hung inside or outside, ensure that the basket is the correct height to fully admire the trailing effect—no one likes to look at the compost-filled top of the basket. Baskets should be hung from a sturdy bracket or pergola, never just from a hook, as they can be very heavy. Remember to

For a fragrant hanging basket, basil and thyme offer the perfect pairing.

water them regularly during warm weather as they dry out more rapidly than plants in the garden.

Plant lavender, chamomile and peppermint in a hanging basket and place it by the barbecue area, or by that special place you sit on a warm, balmy night. The aroma will help repel insects. Similarly, a basket of pretty herbs suspended near the kitchen sink is a most practical decorative device. Lemon balm, thyme and sweet basil will all grow well indoors. For an elegant effect, plant Ground Ivy (*Nepeta hederacea*) around the edges of the basket to balance the herbs in the middle of the basket.

A collection of hanging baskets and tubs can make a small patio or balcony a stunningly colorful and fragrant retreat. They also provide a practical 'garden' for the mobile flat-dweller who could well move on to another home every year.

## Growing In Tubs

Terracotta tubs or large pots filled with scented plants make a lovely feature on patios and balconies. Wooden barrels, cut in half and treated with wood preservative, can also be used.

One of the prettiest gardens I have ever seen was in the middle of the city. Set in a very small courtyard, it was paved with old flagstones, and every available space was filled with oaken tubs that were planted with aromatic herbs and flowers. Silver cotton lavender, balm, and bright geraniums were all colorful and most of the plants had been chosen to retain color throughout the year in the face of harsh, polluted conditions.

Wooden tubs should be painted inside and out with creosote. If the tub has iron bands about it, these need to be treated with a rust preventive; traditionally they are then painted glossy black. Before you fill the tub with soil, place it on bricks or half-shards so that water does not accumulate about the base of the tub. Drilling a few narrow holes in the base of the tub with an awl will facilitate water leaving the tub rather than creating a mushy soil.

The area where the tubs are to be displayed should be sunny; few scented herbs and flowers do well in shade. Be sparing when you plant the tubs—do not be tempted to overfill them with plants, as the plants will grow rapidly. If they are set too close together they will crowd each other for the sunlight. Two to three plants per tub is a fair ratio to use. Ideally, tubs may be planted with small trees, such as the Sweet Bay (*Laurus nobilis*), which may be clipped into a decorative 'ball' shape and underplanted with a complementary herb. Another idea is to plant a mass display of spring bulbs, such as the richly blue Grape Hyacinths.

When filling the tubs, place a few broken pottery shards over the drainage holes so water does not flood out. A little powdered lime is usually advantageous in city-based tub gardens. I was once told that inner city garden soil contains far too much sulfur and soot—a legacy from times gone by—so it is better to ship in fresh soil for tub-filling, rather than take it from the existing garden.

When planting the tubs, try to arrange plants so that their colors contrast attractively, for example, set the crimson, shaggy-flowered plants of bergamot in the center of a tub and surround them with plantlets of golden marjoram as edging. Or, try a selection of different varieties of the same species, such as the sages. The interesting rosy flowers of *Salvia grahamii* look terrific next to the purple leaves of *Salvia officinalis* 'Purpurascens'. Some of the different colored varieties of lemon thyme, such as 'Silver Queen', which has dainty filigree-like foliage, are striking when set with regular thyme.

A wide variety of types of containers will also add interest to a 'tub garden' on a balcony or in a courtyard. Let your imagination have a free rein and experiment. Try a collection of glazed and unglazed earthenware crocks, ceramic pots of different shapes or sizes or even a group of upended chimney pots. The trick to creating unity in such a dissimilar group is to plant the containers with

▸ Lavender hedges can create an intoxicating entrance way.

Two forms of pelargonium make an attractive arrangement for a terracotta pot: dwarf pelargoniums trail over the rim and a taller variety provides the centerpiece.

similar, complementary plants. Scented and ivy-leaf pelargoniums, for example, will always create a pleasing group from disparate containers.

A lovely display may be achieved by massing scented liliums in different tubs and placing them in an appropriate position on a balcony or in a courtyard. A combination of Madonna Lilies and the glamorous Eastern Lilies, for example, would look very exotic in a group of Oriental-style glazed urns or ornamental metal troughs.

If you are fortunate enough to have or acquire an old wooden wheelbarrow, this is an ideal 'tub' for a cottage-style garden. Try planting it with an abundant display of trailing summer bedding plants, such as ten-week stocks, dianthus and candytuft (*Iberis amara*). Do not set plants directly into an old metal wheelbarrow, however; it is better to create a changing display in this type of 'tub', using individually potted plants.

## A Traditional Herb 'Wheel'

As well as being of practical use in the culinary and cosmetic arts, many herbs are rich in fragrance. An old-fashioned way to enjoy these aromatic plants in a garden is to plant them in a circle approximately 6ft/1.8m in diameter, and set rows fanning out from the center to the rim. This idea is derived from the old custom of using abandoned cartwheels where they lay in cottage gardens as kitchen herb garden 'planners'. It is also a most useful device where space is limited.

You can also use different colored varieties of the same herb in the adjoining spaces; for example, green followed by gold marjoram, golden thyme next to pink. Plan your fragrant 'wheel' in the center of a stretch of lawn, preferably where the sun will catch it. Make a 'rim' and 'spokes' of bricks, then plant each segment with a different scented herb or flower. Running a small gravel path around the outside of your wheel will facilitate harvesting, as well as accommodate the idle stroll.

A herb wheel is an ideal way for containing the often vigorous growth of a selection of mints. Plantlets of apple mint, peppermint, spearmint and pennyroyal may be set with bergamot and southernwood, providing a clean refreshing bouquet of scents that will appeal to bees and humans alike. A cottage garden is not complete without the nostalgic scent of certain herbs; clary sage, sweet cicely and borage bring poetry and history to a garden, along with scent and decorative appeal.

Enliven a hedge border with an arched entrance.

## Scented Stairways

The Victorians used scented pelargoniums wherever possible, planting them in protected walled gardens and also bringing potted specimens inside and placing them about walls or in narrow passageways. The leaves would be regularly brushed against by the wide Victorian crinolines of the ladies of the house, so releasing the spicy fragrance for all in the home to enjoy.

A favorite setting for potted scented pelargoniums was up the sides of staircases. Perhaps the best known is *Pelargonium capitatum,* now cultivated for its essence, which replaces rose attar in many perfumes. If you like spicy aromas, the lemony *P. crispum* varieties will suit. They are excellent companions to the oak-leaved geranium (*Pelargonium quercifolium*), which has an incense-like smell, perfect for a winter-time potpourri.

The elegant *Hoya carnosa,* or Wax Plant, is an extremely decorative, scented climber, which is ideal for a hanging basket or for greenhouse cultivation. Other scented plants that bring fragrance to different rooms of the house are cyclamens, primulas and hyacinths.

AN ALPHABET OF

# Perfumed Plants

## *Abelia*

***Caprifoliaceae*** (Honeysuckle family).

***Scent:*** Very sweet and delicate.

***Description:*** A group of evergreen or deciduous shrubs, mostly native to China and Mexico. Softly-shaped, graceful plants, abelias have dark green/gold foliage and clusters of flowers that bloom freely during spring and summer.

***Species:*** The most common varieties are *A.* x *grandiflora* (clusters of white tubular flowers; 6ft/1.8m), which is a popular hedging plant; *A. chinensis* (pairs of rose-tinted flowers; 4–5ft/1.2–1.5m); and *A. floribunda* (drooping clusters of rosy-mauve flowers; 3ft/1m). Other varieties include *A. triflora* (pale yellow flowers with pink or purplish tint; 10ft/3m), which has a soft vanilla scent.

***Cultivation:*** Abelias prefer a warm, sheltered position and will take full sun; against a sheltered wall is ideal, or in shrub borders. A well-drained, loamy soil is required. Moderate selective pruning after flowering to trim and shape will stimulate new growth. Vigorous 'runaway' shoots can be cut right back without harming the plant. Always prune just above a joint shoot.

◂ Climbing roses can be trained to creep attractively over a corner wall.

## *Acacia*

MIMOSA/WATTLE

***Fabaceae*** (Wattle family).

***Scent:*** Penetrating, violet-like and sweet.

***Description:*** Popularly known as wattles, the acacias are widespread throughout Australia, Tasmania, South Africa and the South of France. A national emblem of Australia, it is featured on coins and the official coat of arms.

***Species:*** The most popular species is *A. baileyana* or Cootamundra Wattle (golden fluffy blooms; 10–30ft/3–10m), which is found in many suburban back yards around Australia. *A. dealbata* or Silver Wattle (creamy lemon flowers; 25–100ft/8–30m), is also the 'mimosa' beloved of French street vendors and florists, who ship great quantities to the north of Europe for use as cut flowers during midwinter. It forms a tall, spreading tree and has very pretty, ferny foliage. In mid-nineteenth century France, potted mimosa trees

*Acacia* sp.

were actually brought indoors so their fragrance might be enjoyed during the winter months. Australian children call *A. acuminata* (deep yellow flower spikes; 10–15ft/4–4.5m) the 'raspberry jam tree' because of the sweet smell exuded by the sap when a twig is broken. *A. farnesiana* (tiny golden blooms; 6ft/1.8m) derives its name from the Farnese palace in Rome, having been widely cultivated at Italian flower farms for its cassia oil which, in turn, is used to produce violet perfume types. *A. longifolia*, or Golden Wattle (large bright yellow, ball-shaped flowers; 30ft/10m) is featured on the Australian heraldic crest. Other scented acacias include *A. suaveolens* or Sweet Wattle (pale yellow flowers; 6ft/1.8m), which thrives in seaside gardens and *A. mearnsii*, the Black Wattle.

***Cultivation:*** Acacias require a warm; dry climate and are able to withstand drought. In cooler climates they need to be planted in a sheltered position, such as a high-walled garden or conservatory. A well-drained soil is essential. Acacias may be propagated from seed: like many Australian native plants they have pods that burst open when ripe, scattering the seed over a large area. Collect pods that have not yet burst and store them in paper bags until they open, then plant seeds in seedling trays.

### MIMOSA FRUIT DELIGHT

Complement a selection of fresh summer fruit with the warm fragrance and rich color of mimosa.

*4–6 sprigs mimosa*
*kiwi fruit, sliced*
*rockmelon (cantaloupe) or honeydew melon, chopped*
*seedless green grapes*
*gooseberries*
*blueberries*
*currants*
*mango, sliced*

*Toss salad with Kirsch and a little fresh lime juice. Top with sweetened whipped cream and serve.*

*Achillea* sp.

## Achillea

YARROW

*I rose early in the morning yesterday,*
*I plucked yarrow for the horoscope of thy tale*
*In the hope that I might see the desire of my heart*
*Ochone! There was seen her back to me.*

*An old Hebridean song*

***Asteraceae*** (Sunflower family).

***Scent:*** Pungent, refreshing to smell, reminiscent of feverfew herb.

***Description:*** A large assortment of herbs, sharp in taste, with daisy-like flowers in summer and soft, pretty, silvery foliage, which has a penetrating odor when it is brushed against or crushed in the fingers. It is named for the Greek god Achilles, who was taught of the plant's medicinal benefits by the centaur, Chiron. Yarrow is native to the British Isles and was much used by medieval 'herb-wyfes' for its healing effects. Yarrow tea was drunk to soothe a sore throat and it was often referred to in witches' incantations. In France it was known as 'l'herbe aux charpentiers' because it was used to heal wounds caused by the carpenters' tools.

***Species:*** *Achillea millefolium* or milfoil (very pale pink, tiny flowers and fern-like leaves; 2ft/60cm) is popular as a border plant providing a mat of lush evergreen leaves. Its varieties 'Cerise Queen' and 'Fire King' are especially pretty. The flowers have a soft scent but *A. millefolium*'s chief beauty is its feathery, deep-cut foliage. These leaves were once used in bridal wreaths, giving rise to the country name of 'Venus' tree'. Also pretty is the dainty hybrid *A. 'Moonshine'* (lemon flowers; 6in/15cm), a charming choice for summer bouquets.

***Cultivation:*** Yarrow is a hardy plant, which thrives in a sunny dry position, and prefers a well-drained, sandy soil. Setting plants against a warm brick wall or where they can edge a paved path will enhance the silver foliage and bring out the fragrance. Yarrow may be propagated via root division in spring or autumn; alternatively, take cuttings from established plants in late summer and strike them in a planter box, then plant out the following spring. Yarrow's bright golden flower heads and silvery leaves retain their color, shape and scent very well when dried. They are ideal to harvest for potpourri and dried flower arrangements.

## *Acorus*

SWEET FLAG, SWEET RUSH

TO MAKE A SWEET BAG FOR LINEN

*Take of Orris roots, Sweet Calamus, cypress-roots, of dried lemon-peel, and dried Orange-peel; of each a pound; a peck of dried roses; make all these into a gross powder; coriander-seed four ounces, nutmegs an ounce and a half, an ounce of cloves; make all these into a fine powder and mix with the other; add musk and ambergris; then take four large handfuls of lavender flowers dried and rubbed; of sweet-marjoram, orange-leaves and young walnut-leaves, of each a handful, all dried and rubb'd; Mix all together, with some bits of cotton perfum'd with essences and put it up into silk bags to lay upon your Linnen.*

E. Smith, *The Compleat Housewife*, 1736

***Araceae*** (Arum lily).

***Scent:*** Both leaves and roots have a strong, cinnamon-like aroma.

***Description:*** A group of semi-aquatic herbaceous plants with flag-like leaves. All parts of the plant, flowers, leaves and roots, have a distinctive and spicy fragrance. Also known as 'cinnamon iris' and 'calamus', these plants were much in demand in medieval times for strewing in churches and homes to combat the unpleasant mustiness caused by damp earthen floors and poky windows. There is a record of payment as early as 1516 for flowers and rushes for Henry VIII's chambers. In fact, 'good King Hal' criticized Cardinal Wolsey for being overly lavish with this expensive commodity. The dried roots were placed between linen to scent clothes or powdered for use in sachets or as a paste for cassolettes (an old-fashioned snuff box or jeweled ball, usually attached to a belt or bracelet). Scented rushes are native to Britain and parts of Europe and may also be found growing wild in North America and India. The narrow, iris-like leaves are distilled to produce a volatile oil used in the manufacture of perfumery and some food products. The drug Calamus is derived from the rootstock and the roots are also an ingredient in a well-known fragrance type, chypre.

***Species:*** *Acorus calamus* (yellow and brown flower spikes and upright bright green leaves; 2–3ft/60–90cm). Another variety with creamy white and green variegated leaves, *Acorus calamus* 'Variegatus', which is very striking, is available at most nurseries or may be purchased from a specialist supplier.

***Cultivation:*** In the wild, sweet rushes grow by ponds and streams; in a garden they will do best when planted by the edge of an ornamental pool in 3–5in/7.5–13cm of water. They are unlikely to flower in a garden, but will make a pleasing backdrop for other small, scented, water-loving plants. Plants may be propagated by dividing rhizomes and replanting by the waterside.

## *Albizia*

SILK TREE

***Fabaceae*** (Wattle family).

***Scent:*** Soft, sweet pea-like fragrance.

***Description:*** A group of small, deciduous trees, closely related to the acacia. Albizia are native to parts of Australia, Asia and South Africa. They were originally introduced to Europe from China and seed was transported overland via 'The Silk

Road', hence the derivation of its common name, 'Silk Tree'. Albizia are noted for their softly pretty, fern-like foliage and the delicate, sweet pea-like scent of their clusters of flowers.

***Species:*** Choose either *Albizia julibrissin* (white flowers with distinctive pink stamens; 30ft/10m), which is a dainty tree for a summer show or *A. distachya* (sulfur yellow flowers in bottlebrush-like racemes).

***Cultivation:*** Albizia may be propagated from seed in spring. They are quite fast-growing trees, thriving in full sun, but will not tolerate frost.

### A SWEET-SCENTED PAJAMA CASE

When making a case for storing your night attire, fill the space between the lining and the outer case with:

*1oz/25g lavender flowers*
*1oz/25g thyme*
*1½oz/40g rosemary*
*½oz/15g cloves, crushed*
*1 tspn ground nutmeg*
*2 tspns orris root powder*

*This will perfume your sheets and enfold you in a delicate cloud of scent all night long. Choose pretty cotton or silk to match the bed linen and edge the case with satin ribbon or lace.*

## *Alyssum*

SWEET ALICE OR SWEET ALYSON

***Brassicaceae*** (Mustard family).

***Scent:*** Sweet, honeyed and refreshing, rather like newly mown grass or hay.

***Description:*** A group of many different species, distributed widely throughout the world. A useful,

*Albizia julibrissin*

*Amaryllis belladonna*

pretty and hardy little annual, alyssum is everyone's favorite when it flowers abundantly through spring and summer. The *Planter un Blomen* gardens in Hamburg, renowned throughout the world, are liberally planted with drifts of alyssum and, in summer, tourists flock to the almost overwhelming scent. Alyssum is of a spreading habit. Unfortunately the colored varieties are not as fragrant as the white or pale mauve varieties. They all provide an attractive contrast when interplanted with brightly colored nasturtiums, chrysanthemums or begonias. Being a tiny plant, they are also ideal in pots, window boxes or as an edging plant for paths, gateways and flower beds. Alyssum has the bonus of being especially loved by bees.

***Species:*** *Lobularia maritima* (clusters of white or mauve-tinted flowers; 6in/15cm). Recently developed American varieties include 'Rosie O'Day', 'Pink Heather', 'Lilac Queen' and the purple-flowered 'Royal Carpet'. The smallest cultivar is 'Little Dorrit' (white flowers, 4in/10cm) and the tallest is 'Sweet White'.

***Cultivation:*** Alyssum will prefer a warm, sunny position though they will take a small amount of shade. To propagate, sow seeds in shallow layers in seedboxes, then prick out seedlings during summer and transplant into garden. Alyssum self-seeds rapidly and will soon be found in the most unexpected places within the garden.

## *Amaryllis*

BELLADONNA LILY OR NAKED LADY

***Amaryllidaceae*** (Amaryllis family).

***Scent:*** Powerful and sweet.

***Description:*** A hardy bulbous plant with beautiful clusters of trumpet-shaped flowers each autumn. Amaryllis earned its country name of Naked Lady from its habit of producing leaves that die back before the flowers open on their 18in/45cm long stalks.

***Species:*** This genus contains the single species *Amaryllis belladonna* (pink to rosy red flowers; 2ft/60cm).

***Cultivation:*** Amaryllis should be planted in well-drained soil in a sunny sheltered position. As with most bulbous plant types, they can be a little temperamental about flowering and should not be disturbed unnecessarily. Plant the bulbs, with the necks showing, about 6in/15cm below soil level. When watering, try to imitate the conditions they prefer—dryish summer followed by autumn rains to encourage flowering—and give a light top dressing of decayed manure each winter.

## *Angelica*

TO CANDY ANGELICA

*Boil the stalks of Angelica in Water till they are tender, then peel them and put them in other warm water and cover them. Let them stand over a gentle Fire till they become very green; then lay them on a cloth to dry; take their weight in fine Sugar with a little Rose-water and boil it to a Candy height. Then put in your Angelica and boil them up quick; then take them out and dry them for use.*

From *The Receipt Book of John Nott,* Cook to the Duke of Bolton, 1723

***Apiaceae*** (Parsley family).

***Scent:*** Stalks and seeds have a musky aroma.

***Description:*** A large family of more than 150 species, characterized by glossy green or purple leaves and white flowers, which are fly-pollinated. Angelica also attracts bees to a garden. John Parkinson, in *The Theatre of Plants* (1640) wrote that ' . . . the whole plante, both leafe, roote and seede, is of an excellent comfortable sent, savour and taste', adding, cryptically, that a powder made from the dried roots ' . . . will abate the rage of lust in young persons'. John Gerard wrote that angelica was an anti-witch herb and both the seeds and musky gum from the stems were used in potions to safeguard property and persons. A statuesque perennial plant, featuring strong stems rising from a rosette of leaves, angelica is native to the British Isles and northern Europe. Among Laplanders, angelica was held in high esteem and was used to crown poets so they might be inspired by its scent. According to tradition, angelica's medicinal prowess and ability to ward off pestilence were revealed to humans by the angels, hence the derivation of its name. During the Great Plague (1665), dried and powdered angelica root was mixed with vinegar and used to wash clothes and linen. The seeds were burnt in a chafing dish to perfume and fumigate the home.

Angelica stalks may be blanched and eaten like celery, used in making preserves or candied and used to decorate confectionery. An aromatic oil is distilled from the seeds and roots of angelica and this is used in perfumery and the manufacture of liqueurs, such as Chartreuse. The leaves and roots are used in the production of vermouth.

***Species:*** *Angelica sylvestris* (shiny stalks with a 'bloom' and white flowers; 6ft/1.8m) is the wild variety and *A. archangelica* (fluted shiny stalks with yellowish-greeny flowers; 6.5ft/2m) are the best known forms. Both plants are perennial when cut back and biennial if allowed to flower and go to seed.

***Cultivation:*** Angelica likes moisture but it is hardy and will adapt to most soils. Ideally, it should be planted to the rear of a herb border for its height and large glossy leaves will create an attractive, and protective, backdrop. Angelica plants may be propagated from seed planted at the end of summer when it is ripe, but do not attempt to store angelica seed as it seems to lose its viability quite quickly. When seedlings are a few inches high, thin them to at least 3ft/1m apart and replant in garden, taking care not to disturb the tap root. The plant will flower in the following summer.

## *Anthemis*

CHAMOMILE

TO MAKE OYLE OF CAMOMILE

*Take Oyle a pinte and a halfe and three ounces of camomile flowers dryed one day after they be gathered. Then put the oyle and the flowers in a glasse and stop the mouth close and set it into the sun by the space of forty days.*

*The Good Housewife's Handbook*, 1588

***Asteraceae*** (Sunflower family).

***Scent:*** Refreshing and pungent; when crushed or bruised, the leaves release a fruity, apple-like scent.

***Description:*** A large group of perennial herbs with gray-green, scented, finely serrated leaves and tiny white aromatic daisy-like flowers, native to Europe and parts of America and Asia. Dried chamomile flowers were an ingredient in restorative herbal tonics and tisanes to soothe the nerves and induce sleep. *Ram's Little Dodoen* of 1606 advised that '. . . to comfort the braine, smel to camomill . . . wash measurably, sleep reasonably and delight to heare melody and singing.' 'Anthemis' is derived from the Greek for 'earth apple', referring to the fruity scent released when it is brushed against or sat upon. Chamomile is noted for its creeping habit, the stems lie flat and send off roots as they go along, so it has long been used to make aromatic lawns. According to Shakespeare, Falstaff—who said of chamomile that 'the more it is walked upon, the faster it grows'—was very fond of walking upon his own chamomile lawn and it is likely that Sir Francis Drake's famous game of bowls was played on a chamomile lawn. In days gone by, the stone seats in 'herbers' (arbors) were covered with chamomile and it was used in

▸ '. . . though the leaves of the [lily] flower be white, yet within shineth the likeness of gold . . .'
The stately beauty of Arum lilies.

*Anthemis* sp.

lieu of turf or gravel for paths. John Evelyn wrote of the arrival of cooler weather in the Northern Hemisphere, saying '. . . in October it will now be good to Beat Roll and Mow carpet walks and camomile, for now the ground is supple and will even all inequalities.' Folklore states that chamomile is 'the plants' doctor' as it will supposedly benefit a sickly plant if it is set nearby. The dried leaves may be used as a tobacco substitute, while a cooled infusion is a most effective scalp tonic.

***Species:*** *Anthemis nobilis* (daisy-like white flowers; 12in/30cm) is the variety most often used for lawns. Other choices include *A cupaniana,* the Rock Chamomile (white flowers and pretty gray foliage; 14in/35cm), which makes a delicious, scented tea; *A. arvensis* or Corn Chamomile, which has a chrysanthemum-like scent; or double-flowered *A nobilis* 'Plena'. *A nobilis* 'Treneague' is also much favored for lawn-making. However, be sure to avoid *A. cotula,* or Stinking Chamomile, which has a dreadfully rank smell and it will also blister the skin of the unwary.

***Cultivation:*** Chamomile will become established in most types of soil and prefers a reasonably sunny position. Plants must be well-watered in summer and protected from frost each winter. Chamomile may be propagated directly from seed, or seedlings may be raised in planter boxes before being placed in the garden. To make a chamomile lawn, plant seedlings 6–8in/15–20cm apart in spring. Keep soil moist and free from weeds for the first year, by which time they will have formed a dense mat that will suppress all weeds that follow. After the first year chamomile plants may be lightly clipped to encourage them to spread; after the second year, the lawn may be mowed, but only on a very high setting so the plants are not damaged.

### CHAMOMILE FLOWER BALM

Rinsing the complexion with 'swete waters', especially those made from rain or spring water, has long been believed to beautify the skin. Such waters were once kept in elegant silver ewers in ladies' bedchambers. Use this freshening lotion to whisk away the last traces of makeup, soap and grime which can dull the skin.

*2 tbspns strong, warm chamomile infusion*
*1 tbspn strong, warm elderflower infusion*
*3½fl oz/100ml rosewater*
*2 tspns witch hazel*

*Place all the ingredients in a lidded glass jar or bottle and shake well before applying with a moistened cotton ball.*

## *Anthericum*

St Bernard's Lily or St Bruno's Lily

*Nothing is more gracious than the lily in fairness of colour, in sweetness of smell and in effect or working and vertue.*

*Bartholomaeus Anglicus*, fifteenth century

*Anthericum larrendenii liliago*

***Liliaceae*** (Lily family).

***Scent:*** Sweet, cosmetic, powdery quality.

***Description:*** A group of about 40 species with tuberous roots found in parts of South Africa, South America and the Central European Alps. The pure white starry flowers are tipped with green and have a strong sweet scent. They are particularly attractive when planted in clumps in the garden. The slender stems and fleshy roots have no fragrance, although they were once harvested as an ingredient for pomades to remove wrinkles and whiten the complexion. The spidery-looking flowers were once, according to the lore of the Doctrine of Signatures (a belief that plants signified certain applications and uses by their appearance), deemed a cure for the bite of a poisonous spider—hence the plant's common or country name, Spiderwort.

***Species:*** *Anthericum liliago* (large white funnel-shaped flowers with yellow stamens; 2ft/60cm) is closely related to the Alpine Asphodel in appearance. It is hardy and will flower from early summer right through to autumn, filling the air with its unique perfume.

***Cultivation:*** Anthericum tubers should be set with their tips only an inch or two below the surface. They prefer a sunny position and will greatly appreciate a mulch around their basal leaves to keep roots cool in summer and to protect the plant during a harsh winter. Most lilies also need a companion plant, for example, clary sage, to protect their ripening stems. Anthericum lilies flourish in the classical 'cottage garden' setting, perhaps because the well-cultivated soil in older gardens is usually alkaline and rich in wood mulch. If your soil is overly acid, turn limestone chips and sand through it to redress the balance.

## *Aponogeton*

Water Hawthorn

***Aponogetonaceae*** (Pondweed family).

***Scent:*** Like almond or vanilla, depending on the species.

***Description:*** A genus of semi-aquatic plants with creeping roots and floating or partially submerged leaves. Native to South Africa, where it has the common name of Cape Asparagus (the flower heads were once harvested and eaten rather like young asparagus tips), water hawthorn is also found widely in Australia. It may also be grown in cooler climates, provided it is given a fairly well-protected position. Water hawthorns have narrow, oval, light green leaves with maroon tracings and richly scented snow-white flowers. Unlike other pondweed plants, these are pollinated above the water's surface by insects, rather than below. They are a very pretty choice for an outside pond or

*Aponogeton distachyos*

decorative tub in a greenhouse and, once established, can live for up to 50 years.

***Species:*** *Aponogeton distachyos* (deeply lobed white flowers with black anthers; vanilla scented) will flower from spring through to autumn.

***Cultivation:*** Plant the rhizomes in mud 8–18in/20–45cm below water surface so leaves lie flat on water in full sun or semi shade. (NOTE: full sun is required for flowering but ensure plants do not 'bake' in the water.) There needs to be a sufficient volume of water to keep the temperature level. Allow at least 2ft/60cm between clumps of rhizomes as the leaves will spread enthusiastically. Water hawthorn will not grow in stagnant water or infertile soil. A fresh water current is required in its natural state, so if planting them in an ornamental pond, invest in an efficient water purification and circulation system. If growing water hawthorn in greenhouse tubs, flush them with pure water daily to remove dust and grit. Keep an eye out for water snails; although they are worthy consorts for most aquatic plants, helping to clear leaves of scum, they have a cannibalistic attitude to water hawthorn and will decimate the young shoots of a plant.

SCENTED SLEEPY PILLOW

*1 cup sweet woodruff*
*1 cup rose petals*
*½ cup southernwood*
*3 bay leaves, crumbled*
*2 tspns cloves, crushed*
*2 tspns rosemary*
*5 drops rose oil*
*small cotton or muslin pillowcase*

*Crush all the herbs and combine with rose oil in a china bowl. Spoon the mixture into the pillowcase and stitch or tie securely. Tuck into a pretty pillowcase or slip under a regular pillow where the scent will linger all night long.*

## *Arbutus*

STRAWBERRY TREE

***Ericaceae*** (Heather family).

***Scent:*** Soft, sweet, musky, honey fragrance.

***Description:*** A handsome group of evergreen small compact trees, renowned for their delicious, strawberry-like fruits. The flowers are also exceptionally pretty, being of a creamy color, heavily scented and shaped like little lily of the valley bells. The trees bloom early in spring, and are replaced by fruit for most of the summer months. In early Rome, Pliny named the Strawberry Tree *Arbutus unedo*—'unedo' being a corruption of *unum edo* meaning the fruit was so delicious the picker should avoid being greedy and take only one. In later times, the strawberry tree was discovered in America and its seed was introduced to England in the late seventeenth century. Henry Compton, who was then Bishop of London, grew strawberry trees at his garden at Fulham Palace, along with other rare plants from the New World, such as sassafras and hickory trees. Strawberry trees tend to be better suited to cooler climates and are now found throughout Ireland, Greece and Italy as well as America and Britain.

***Species:*** *Arbutus andrachne* (green/white flowers; 10ft/3m) is native to Greece and Crete; *A. unedo* (creamy flowers; 20ft/6m) is the best known variety, sharing with its Californian cousin *A. menziesii* the visual bonus of an attractive, cinnamon-colored bark.

***Cultivation:*** For a strawberry tree to bear fruit, it should be planted by two or three others, as this facilitates pollination. It is a tender tree and should be planted in a sheltered position. An ideal place is in a small, walled garden. It enjoys lime in the soil and requires adequate mulching before fruiting will occur. Propagation is either from seed or cuttings.

*Arbutus andrachne*

## *Artemisia*

SOUTHERNWOOD AND TARRAGON

*The seed of flax put into a Raddish root or sea onion and, so set, doth bring forth this herb Tarragon.*

John Gerard, *The Herball*, 1597

***Asteraceae*** (Sunflower family).

***Scent:*** Pungent, distinctive, refreshing, usually pleasant.

***Description:*** A large group of perennial herbs or small shrubs, rich in eucalyptol and sharp to the taste and smell. They are hairy in appearance with feathery leaves and small drooping flowers that vary in size and shape between species. Since early times all the species have been held in high regard for their medicinal, culinary and household uses.

***Species:*** *Artemisia abrotanum* or southernwood (4ft/1.2m) was once known as Lad's Love or Old Man because it was thought to promote the growth of a new beard. Culpeper wrote: 'The ashes thereof, mingled with old sallet oyle, halps those what have their hair fallen and are bald causing their hair to grow again, either on the head or on the beard.' A native of Europe and Britain, southernwood was

*Artemisia abrotanum*

much used in potpourri, to keep moths at bay and in sleep-pillows. It has a very strong scent, almost pure eucalyptol. Louise Beebe Wilder, in *The Fragrant Path,* wrote that she loved to use southernwood with pink moss rose-buds for special nosegays. *A dracunculus* (2ft/60cm) and *A. dracunculoides* (2ft/60cm)—French and Russian, respectively—are better known as tarragon.

French tarragon is the variety a cook should always plant, being far more flavorsome. It is a particularly appropriate complement to salads, chicken and fish. Originally native to Siberia, tarragon was introduced to France by the Moors. Its name, *dranunculus,* means 'little dragon' because its roots curl back like a dragon's tail or claws. The roots were once used to cure toothaches and neuralgia and John Evelyn wrote that tarragon was ' . . . highly cordial to head, heart and liver'. Its crushed leaves have a strong, absinthe-like scent.

Other varieties include *A. absinthum* (bright orange flowers), which has a sweeter fragrance and is used in the production of absinthe (one of the 'bitter herbs' mentioned in the Bible); Roman wormwood or *A. pontica*; and *A. borealis.*

***Cultivation:*** Seed for artemisia species is, generally, hard to obtain, so propagation is usually by cuttings or from purchased seedlings. Artemisia plants are hardy and prefer a sandy, well-drained soil—the drier the soil, the sharper the fragrance and flavor. They are a good choice for a sturdy low hedge, for borders, or for planting against a sunny wall or rockery. Plants should be pruned hard in spring and again, if necessary, during summer, otherwise they will become 'leggy' and untidy.

### SOUTHERNWOOD HAIR RINSE

Southernwood, once known as Lad's Love, is very good for the hair. An infusion may be dabbed directly on to the scalp to ease irritations, flaking or dandruff. Use this as a final aromatic rinse after shampooing. It will envelop you in a soft cloud of fragrance and also help to offset any scalp dryness or fly-away hair after shampooing.

❖

*3 tbspns southernwood*
*1 tbspn clary sage oil, or essential oil of your choice*
*9fl oz/250ml apple cider vinegar*
*1½fl oz/50ml whisky*

*Combine all the ingredients in a lidded jar and steep for two weeks. Strain and re-bottle.*

## *Baeckea*

AUSTRALIAN MYRTLE

***Myrtaceae*** (Myrtle family).

***Scent:*** Strong, camphoraceous.

***Description:*** A group of small trees or shrubs native to Australia and some nearby islands. They are powerfully aromatic evergreens with leaves that emit a camphoraceous fragrance and flowers that appear in late summer and early autumn. Aborigines in Australia would crush the leaves and use them as a medicine to clear the chest or nasal congestion.

*Baeckea ramosissima*

***Species:*** *Baeckea camphorosmae* (pale pink flowers; 3ft/1m) is an erect shrub, while *B. gunniana* has a prostrate habit; *B. ramosissima* has pink to white flowers.

***Cultivation:*** Baeckea are quite hardy plants, provided the night temperature does not fall beneath 48–50°F/9–10°C as they are susceptible to frost. They prefer a heavy soil and a protected, sunny position—an ideal spot might be under an arbor or by a path, where the leaves are regularly brushed against, thus releasing their refreshing aroma.

## *Bauhinia*

ORCHID TREE, BUTTERFLY TREE

***Fabaceae*** (Cassia family).

***Scent:*** Delicate, vanilla-like.

***Description:*** A small group of attractive, ornamental, medium-sized deciduous trees or large shrubs with pretty twin-lobed leaves and orchid-like flowers in white, pale pink and dark purple. The plants were named after the sixteenth century herbalists, Jean and Caspar Bauhin. At the time, eminent botanist Carl Linnaeus noted that '. . . the two lobed leaves recalled the noble pair of brothers'. Bauhinias, native to India, East Africa and Western China, are now a widespread ornamental plant in Australia and other parts of the world. Also known as the 'buddhist tree', it is considered sacred to Buddha and is a regular feature in temple gardens. For many years, a decoction of the bark was used to treat skin problems and the pickled flowers were a popular, if rather sour, condiment. In India, the aromatic leaves are dried and used in producing fragrant cigarettes.

***Species:*** *Bauhinia purpurea* (pink/purple flowers; 20ft/6m) or Purple Bauhinia is the best known variety, providing a striking picture in many suburban gardens and flowering right through a temperate winter into spring; also *B. variegata* (mauve flowers; 20ft/6m).

***Cultivation:*** Bauhinias are a most attractive ornamental tree in a garden, provided they are placed in a warm and sheltered position. Once established they are quite hardy and will flourish for many years. Adequate mulching is essential to protect the root system. Propagation is via cuttings.

## *Billbergia*

***Bromeliaceae*** (Bromeliad family).

***Scent:*** Soft and sweet.

***Description:*** A group of epiphytic plants with thin, strappy leaves and exotically fragrant flowers. Billbergias grow best when attached to other

*Billbergia nutans*

trees—they take in the air through their roots and do not need nutrients found in the soil. Some varieties, however, may be potted in very light compost and taken indoors so their scent may be enjoyed. Billbergias are native to the rainforests of South America and Central America and will grow in most humid and tropical areas. In Brazil, they are tied to balconies of elegant townhouses and their fragrance is enjoyed by the inhabitants of the apartment and the pedestrians below.

***Species:*** The key variety is *Billbergia nutans,* also known as 'Queen's Tear Drops' because its yellow and red flowers look like droplets.

***Cultivation:*** As it is an epiphytic plant, a billbergia should be strapped firmly to the trunk of another tropical or semi-tropical tree, in semi-shade for best results. If the plant is to be potted, use a compost made up of equal parts of sand, peat and loam and pinch off old shoots regularly. When the plant has stopped flowering, it must be repotted in fresh compost.

*Borago officinalis*

## *Borago*

BORAGE

*The vertue of the conserve of borage is especially good against melancholie; it maketh one merie.*

John Partridge, *The Treasurie of Commodious Conceits and Hidden Secrets,* 1586

***Boraginaceae*** (Borage family).

***Scent:*** Cooling, fresh, cucumberish.

***Description:*** A group of hairy-leaved annuals with drooping bright blue flowers. Native to the Mediterranean, the early Romans first introduced borage to Britain where its leaves were used in salads before lettuce became popular. The flowers were candied, much as rose petals are today, and used in confectionery. According to the mysterious 'W.M.', cook to Queen Henrietta Maria in 1655, a 'Conserve of Borage flowers after the Italian Manner' could be prepared thus: 'Take of fresh Borage flowers four ounces, fine Sugar twelve ounces, beat them well together in a stone mortar, and keep them in a vessel well glazed.' Borage's country name of 'Cool Tankard' refers to its use in scenting wine cups, iced tea, tonics and cider. It was also a valued pot-herb being used to give 'soop . . . an excellent cordial savour'.

Traditionally borage was believed to impart bravery and its name is thought to have come from the old Celtic *borrach,* meaning 'courage'. Borage tea was drunk by knights before they competed in jousts and their ladies would give them kerchiefs embroidered with the blue flower motif for luck. Dr Fernie in his *Herbal Simples* (1895) attributed borage's invigorating effect to the fact that ' . . . the fresh juice contains 30 per cent nitrate of potash'.

***Species:*** The key species is *Borago officinalis* (sky blue flowers; 18in/45cm tall). Pliny called it *euphrosynum* because it 'maketh a man joyfull'

▶ Arc-shaped beds in a circle of lawn are interplanted with iceberg roses and white viola.

and, along with the violet and the rose, it was a 'cordial flower for cheering the heart'.

***Cultivation:*** Borage seed should be sown in a sunny position in a light, well-drained soil in early spring. The plants should later be thinned out. If left alone, it will readily self-seed. To harvest flowers for potpourri, pick them just before they fully expand, otherwise they will lose their color. Dry them immediately on wire racks and store in a cool dry place.

Prune borage plants back quite hard after flowering to remove all dead leaves and stems. An ideal spot for borage is in a flower bed with ornamental shrubs as the bright blue flowers create an appealing contrast.

### BORAGE SORBET

❖

*6oz/175g sugar*
*7fl oz/200ml water*
*7fl oz/200ml champagne*
*1 cup borage flowers*
*juice of 1 lemon*
*2 large (2fl oz/60g) egg whites, whisked*

*Dissolve the sugar in the water and champagne. Bring to the boil and infuse borage flowers for 30 minutes. Add lemon juice, strain the mixture and pour it into a metal freezer tray. Chill until the mixture begins to freeze, then fold through egg white(s). Freeze until firm.*

## Boronia

***Rutaceae*** (Rue family).

***Scent:*** Sweet, fresh, lemony.

***Description:*** A large group of heath-like shrubs native to Australia, especially to New South Wales, the boronia's flowers were well-loved by American gardening writer Louise Beebe Wilder who described them as '. . . amongst the sweetest-smelling in the world . . . a combination of violets, lemon-verbena and roses'. Although boronias are commonly grown in gardens, commercial cultivation of the plants for perfumery production has not always been successful. The elusive scent is still more likely to be detected in the Australian bush on a hot day than in a bottle, although boronia is being increasingly farmed in the perfume-making areas of Southern Europe.

*Boronia heterophylla*

Boronias have wispy twigs with narrow, needle-like evergreen leaves which are aromatic when crushed, though the aroma is not as strong as that of the flowers. The tiny, bell-like flowers appear in late winter and bloom right through to mid-summer, scenting the air with their famed lemony-floral fragrance.

***Species:*** *Boronia megastima* (browny-red flowers, yellow within; 1½–3ft/0.5–1m) or Brown Boronia is native to Western Australia. A profusely flowering variety, this is the one grown for perfume production. *B. serrulata* (purplish-pink flowers; 1½–3ft/0.5–1m) is also known as the Native Rose, as it has distinctive, slightly oval and serrated leaves, similar to the leaves of a rose. Also attractive

*Bouvardia* hybrid

are *B. ledifolia* (large bright pink flowers; 1½–3ft/0.5–1m) or Sydney Boronia and *B. pinnata* (dark pink flowers; 3–6ft/1–2m) or Pink Boronia are both free-flowering and highly aromatic. *B. citriodora* (pale creamy pink flowers; 1½–3ft/0.5–1m) or Lemon Scented Boronia is a particularly hardy variety and will tolerate quite a lot of frost.

***Cultivation:*** Boronias require regular watering and a well-drained soil. Adding a little gypsum to the soil is advantageous. Boronias will tolerate a variety of aspects but seem to be happiest in dappled shade in a reasonably sheltered position. In cooler climates, boronia is usually only suitable as a cultivated greenhouse pant. Boronias should be pruned hard after flowering, taking back about a third of the foliage. This will ensure the plants remain fresh and attractively scented.

## *Bouvardia*

***Rubiaceae*** (Madder family).

***Scent:*** Sweet, vanilla-like.

***Description:*** A group of evergreen woody plants, native to Mexico and South America. Also known as the 'flores de San Juan', bouvardia was much favored by the wealthy families of San Diego and Rio de Janeiro who had containers filled with the plants in every room. The pretty, starry florets were also thought a lucky choice for a bridal bouquet or buttonhole. The plant was discovered by Baron Humboldt, whose name is remembered in the Latin name for the white-flowered variety, *Bouvardia longiflora*.

***Species:*** *Bouvardia longiflora* (large white flowers) and *B. jasminiflora* (small jasmine-like flowers) are highly fragrant and make ideal choices for potted specimen plants. *B. ternifolia* (scarlet flowers) has no perfume at all.

***Cultivation:*** Bouvardia requires a warm temperature. If being cultivated in cooler areas, it must be brought inside during winter months and cold evenings—a constant temperature of 65–70°F/18–20°C is preferred. Propagation is via root cuttings. Plant in sandy compost and keep in even temperature in planter tray in a greenhouse. When the plants are established, move them to larger pots. Bouvardia's growth is sprawling and it will need regular tying back to firm stakes. Flowering is luscious, continuing from summer through to autumn; during flowering, pinch out new shoots to avoid legginess. In autumn, after flowering, cut back and reshape properly, using the wood cuttings to propagate new stock.

## *Buddleja*

SUMMER LILAC, BUTTERFLY BUSH

***Loganiaceae*** (Logania family).

***Scent:*** Sweet, musky, honeyed.

***Description:*** A group of very attractive deciduous shrubs or small trees that bring color as well as fragrance to a garden. Buddlejas are native to South America and Asia but, being vigorous growers, are hardy in most cooler areas. The plant was discovered by the quaintly-named eighteenth century botanist Rev. Adam Buddle and named in his honor. The old variety, *Buddleja davidii*, which most people know, was named for Pere Armand David, a French missionary, who discovered it while traveling in China. Buddlejas make an attractive lightly scented hedging plant, particularly if different colored varieties are placed by each other. Each plant bears many fat spikes, each of which carry masses of tiny blooms. These attract many different types of butterflies, hence the derivation of its country name. A border of these brightly colored, fragrant plants is a very pretty sight. Choose the shorter, more recent forms for a border; other buddlejas will quickly grow to 10ft/3m and need regular pruning to retain their shape.

***Species:*** *Buddleja davidii* (white through to dark purple flowers; 10ft/3m) is the best known. This is a graceful tree and its sweeping habit is similar to a willow tree. Several very attractive varieties are available, including ones with violet, almost black flowers, rich scarlet flowers, soft pink flowers, and pure white flower spikes, which are lovely for a white garden plan. These varieties make a splendid show when planted together, in rows or a clump.

Other varieties include the exotic Chilean *B. globosa* (orange, ball-shaped flowers; 12ft/3.5m) and *B. fallowiana* from Western China (pretty pale blue flowers), which also boasts attractive silvery foliage. *B. alternifolia* (pale lilac flowers; 10ft/ 3m) is the best choice for a free-standing tree, as its arching branches lend themselves well to pruning and shaping, creating an elegant shape.

***Cultivation:*** Buddlejas grow best in sandy loam and prefer the additional protection of being planted against a warm sunny wall or fence. They tend to do reasonably well in a seaside position and will thrive when a little lime is added to their soil. Buddlejas may be featured as fragrant trees in a garden or even—for the very patient—trained along wire to create a pleasing shape. It is important that buddlejas are pruned back hard after flowering, otherwise they will become straggly and the blooms will be only small.

## *Buxus*

BOX

*. . . the fragrance of eternity, this is one of the odours which carry us out of time into the abyss of the unbeginning past; if we ever lived on another ball of stone than this, it must be that there was Box growing on it . . .*

Oliver Wendell Holmes

▶ Named for the Greek goddess of the rainbow, the iris is famed for its stately and attractive flowering habit.

***Buxaceae*** (Box Tree family).

***Scent:*** Strong, aromatic, pungent, distinctive.

***Description:*** A group of evergreen shrubs and small trees, box has long been used for ornamental hedging and topiary. It was also cultivated for its strong, heavy wood, which was once much used for making furniture, particularly musical instruments and printers' blocks. Native to Britain and Northern Europe, it is found widely throughout the world and is also strongly identified with the east coast of America. In *Old Time Gardens* (1905) Mrs Earle wrote that the 'unique aroma of Box, cleanly bitter in scent as in taste, is . . . almost hypnotic in its effect. This strange power is not felt by all, nor is it a present sensitory influence; it is an hereditary memory, half known by many, but fixed in its intensity in those of New England birth and descent, true children of the Puritans; to such ones the Box breathes out the very atmosphere of New England's past'.

However, not everyone enjoys the scent of box. Queen Anne had all the box hedges at Hampden Court destroyed because she disliked them so much. In *Adventures in My Garden,* Louise Beebe Wilder wrote: ' . . . the pungent smell of the leaves is to me highly refreshing and stimulating, but all do not like it. Near me is a cottage half surrounded by a fine Boxwood hedge but of it the woman who dwells therein said: "It's gloomy and I don't like the smell all through the day and night".' Box's scent is especially noticeable when the foliage is wet or when the aroma is volatalized by hot sun. The pale yellowish green flowers are small and inconspicuous and appear in spring. They have a soft sweet mignonette-like scent, which is quite different from that of the foliage.

***Species:*** *Buxus sempervirens* is a tall variety with shiny green leathery leaves. It is the best choice for topiary or for a clipped formal hedge or arrangement in a courtyard or doorway. *B. microphylla,* or Little-Leaf Box, is a low growing variety suitable for hedging or dwarf topiary, as is the Dutch Box, or *B. sempervirens* 'suffruticosa'. *B. elegantissima* is very attractive for parterre work or—for the ambitious—an old-fashioned knotte garden, with its silvery foliage trimmed with white.

***Cultivation:*** Box are very hardy plants and will grow in most situations. They prefer a slightly chalky soil. Propagation is by root division or cuttings taken at the end of summer flowering.

## *Calendula*

MARIGOLD

> *The yellow leaves of the flowers are dried and kept throughout Dutchland against winter to put into broths, physicall potions and for divers other purposes, in such quantity that in some grocers or spicesellers are to be found barrells filled with them and retailed by three penny or less, in so much that no broths are well made without dried Marigolds.*
>
> John Gerard, *The Herball,* 1597

***Asteraceae*** (Sunflower family).

***Scent:*** Warm, pungent.

***Description:*** Pot marigolds, frequently called 'golds' in olden times, comprise a genus of about twenty species of hardy flowering annuals. Their old country name, recorded by Shakespeare in *The Winter's Tale,* was 'Mary-buds'. The pot marigold was once taken as a symbol of the Virgin Mary, being thought to bloom in time for certain religious festivals. It was also thought to symbolize constancy as the flowers always turn to face the sun. Shakespeare wrote of this habit that ' . . . the marigold . . . goes to bed wi' the sun and with him rises, weeping.' Pot marigolds are native to Southern Europe and widely grown throughout the world. The botanical name *Calendula* is derived from the Latin *calends,* for pot marigolds were said to bloom promptly on the first day of every month in the old Roman calendar.

A valued pot herb and popular cottage garden plant, pot marigolds have bright green leaves and golden flowers. They grow quickly from seed and will self-sow in successive summers. Vita Sackville-

*Calendula officinalis*

West wrote that you should, 'for Summer's quick delight, sow Marigold . . . ' The culinary and medicinal properties of pot marigolds have long been known. John Parkinson advised pot marigolds be used in 'possets, broths, drinks . . . as a comforter to the heart', and decoctions were once used for treating many ailments, from smallpox and rheumatism to burns and bruises. The flowers and leaves were used by field doctors during the American Civil War as a first aid poultice for cannon-shot wounds sustained by soldiers. All parts of the plant are useful: marigold water is soothing for bathing eyes and skin, marigold essence heals chafed skin, the leaves add tang to soups and stews and the flowers add color to summer salads. In particular, the petals are a pretty addition to rice dishes, adding flavor and a slightly golden hue. On a more frivolous note, pot marigolds were also a popular cosmetic and fashion accessory. Ladies of Elizabeth I's court used a rinse from the flowers to add golden luster to their hair and it is recorded that pot marigolds pickled in vinegar and massaged into teeth and gums were 'a soveraigne remedie for the assuaging of greivious paine of the Teeth'. In *The Proffitable Arte of Gardeninge* (1568), Thomas Hyll wrote that 'This Marigold is a singular kinde of herbe sowen in gardens, as well as for the pot, as for decking of Garlands, beautifying of nose-gays and to be worne in the bosome.'

Pickled marigolds may be made by adapting the following recipe from *The Country Housewife and Lady's Director* (1732); try using them to add zest to salads and cold meat:

❖

To Pickle Marigold Flowers—*Strip the flower-leaves (petals) off, when you have gather'd the flowers, at noon, or in the heat of the Day, and boil some Salt and Water; put your Marigold flower-leaves in a gallypot, and pour the salt and water upon them; then shut them up till you use them and they will be of a fine colour and much fitter for Porridge than those that are dry'd. (Strain before use)*

*Calendula officinalis*

***Species:*** The key species is *Calendula officinalis* (bright yellow or orange flowers with frilled petals; 12–15in/30–40cm high). The color and size of flowers varies; choose from 'Radio' (bright orange), 'Mandarin' (reddish-orange), 'Golden Beam' (rich yellow) or 'Cream Beauty' (very pale yellow). Sow all the different varieties for a glowing display with an attractive tangy aroma to contrast with sweeter-scented plants nearby.

***Cultivation:*** Pot marigolds are quite hardy so seed may be sown directly into a warm sunny place in the garden. Plant in spring, allowing 12in/30cm between plants. Pot marigolds will grow in most soil and will start flowering in early summer, continuing into late autumn. Cutting back some of the early dead heads will result in a second flowering. Pot marigolds will self-sow throughout the garden and spread rapidly, if not thinned out regularly.

### MARIGOLD PANCAKES

❖

*3½oz/100g plain (all-purpose) flour, sifted*
*1 tspn sugar*
*pinch each salt and baking soda*
*10 marigold flowers, finely chopped*
*3½floz/100ml sweet white wine*
*2 apples, peeled and puréed*
*2 egg whites, beaten*
*1½oz/40g butter, melted*

*Combine the flour, sugar, salt, soda and marigolds. Add wine, apple purée and egg whites, stirring carefully to form a smooth batter. Leave for 1 hour, then fold through butter. Fry spoonfuls of batter in a hot skillet, turning rapidly until crisp at the edges. Serve with caster (superfine) sugar and lemon.*

## Carum

CARAWAY

*Come cousin, Silence! Nay you shall see mine orchard where in an arbour we will eat a last Year's pippin of my own graffing with a dish of carraways . . . And then to bed.*

Wm Shakespeare, *Henry IV (II)*

***Apiaceae*** (Parsley family).

***Scent:*** Sweet and slightly spicy.

Flowers of rocket, *Anemone* and *Achillea* blend harmoniously.

***Description:*** Caraway is a biennial plant, primarily cultivated for its aromatic seeds. Native to all parts of Europe, it may be grown in temperate climates throughout the world. Most of the world crop of caraway seed, interestingly, comes from Holland where it is used to flavor their favorite liqueur, Kummel, along with aromatic bitters.

Caraway is well known as a culinary herb; bread, cheese, coffee, hot chocolate and soup may all be flavored with the seeds. Nicholas Culpeper wrote that the young roots were made into 'confects, once only dipped in Sugar' which were 'a most admirable remedy for those that are troubled with wind'. Similarly, caraway oil remains one of the active ingredients in many commercial brands of 'gripe water', used to soothe colic in babies, and other types of medicine. The tender roots were once used as a vegetable, rather like parsnips, and the leaves will add aroma and flavor to sauces.

Much folklore surrounds caraway. Thought to prevent theft and infidelity, seeds were baked into cakes and given to soldiers by their sweethearts at home. A medieval love potion advised girls to plant caraway in soil where their beloved had just trodden and caraway seeds were fed to chickens and geese to stop them wandering.

***Species:*** The key variety is *Carum carvi* (3ft/1m), a tall herb with small green thread-like leaves and soft umbrella shaped clusters of green flowers, which turn to seed in late summer.

***Cultivation:*** Caraway is reasonably hardy and will tolerate light frost. It will grow in most soil types, but prefers slightly acid conditions. Sow in early autumn or late spring, setting seed about a foot apart, and thin the plants when they come up. To harvest the caraway seeds, gather them when they ripen and store in airtight containers.

## *Cassia*

***Fabaceae*** (Cassia family).

***Scent:*** Sweet, vanilla-like.

***Description:*** A group of fast-growing, ornamental trees and shrubs that are native to Eastern countries and to South America. Cassias are very hardy and will grow in most areas. With their evergreen foliage and bright, bell-shaped yellow flowers, they create a great impact in a garden. In South America, cassia trees are planted alongside major arterial roads in the cities.

***Species:*** *Cassia corymbosa* (10ft/3m; profuse golden flowers) is native to South America. The pods of *C. fistulosa* (25ft/8m; rich gold flowers) carry a soft brown pulp, which Bengalis use to flavor tobacco.

***Cultivation:*** Cassias like a well-drained sandy soil, with a little peat or blood and bone worked in. They should be planted in a warm, though not necessarily sunny, position and pruned regularly to ensure plenty of flowers. Cassias are hardy and suited to most temperate or coastal climates, though they are susceptible to frost.

## *Centaurea*

SWEET SULTAN

***Asteraceae*** (Sunflower family).

***Scent:*** Rich, musky and very sweet.

***Description:*** Although there are more than 600 species in this genus, it is only the annuals known as 'Sweet Sultans' that are bestowed with fine fragrance. Native to Turkey, and named in honor of the Sultan of Constantinople, Sweet Sultans were introduced to Paris in the seventeenth century and rapidly gained popularity throughout England and Europe for their bright, powder-puff like flower heads and soft, sweet, meady scent. In his *Paradisi in Sole Paradisus Terrestris* (1629) John Parkinson described Sweet Sultan as '. . . a stranger of much beauty, lately obtained from Constantinople, where the great Turke liked it and wore it himself . . . of so exceeding a sweet scent that it surpasseth the finest civet there is.'

***Species:*** Sweet Sultans are indispensable in a scented garden and their honeyed perfume makes them ideal for cutting. Flower color and plant size vary. Choose from *Centaurea moschata* (white, purple or yellow flowers; 2ft/60cm) or the Giant Sweet Sultan (white flowers; 4ft/1.2m). Massed in a mixed border, one or all of the Sweet Sultans is worth growing. The Common Centaury is widespread throughout England and Europe. A scented oil may be obtained from this plant via distillation but, in the garden, it is quite scentless.

*Centaurea moschata*

***Cultivation:*** Sweet Sultans may be raised from seed. Sow seeds ¼ inch deep in early spring and they will grow quickly, flowering through summer and into early autumn. Adding a little lime to the soil will ensure profuse flowering.

## *Cestrum*

NIGHT JESSAMINE

***Solanaceae*** (Nightshade family).

***Scent:*** Very sweet, fruity.

***Description:*** This subtropical genus is native to South America, the West Indies and the South Pacific Islands. Although the simply-formed, deciduous shrubs are quite plain and unprepossessing, the night-blooming, greenish-yellow or white tubular flowers ensure the inclusion of *Cestrum* plants in any scented garden. The fragrance intensifies as the temperature falls at nighttime, with the star-like flower buds opening fully to release perfume.

***Species:*** *Cestrum parqui* (green/yellow clusters of flowers; 8ft/2.4m) will do best planted near a patio, under a window or against a wall that retains

*Cheiranthus cheiri*

the sun's warmth during the day. *C. nocturnum* (creamy white flowers; 8ft/2.4m) is a pretty choice but the bright orange flowers of *C. aurantiacum* offer more color and impact.

***Cultivation:*** *Cestrum* species are frost-tender and require a moderate climate. They will grow as either tall shrubs or may be trained up a pillar or portico. A well-drained, rich soil in a warm sheltered position is ideal. The tubular flowers appear during midsummer, both from the ends of shoots and the leaf joints. However, judicious watering and use of a good liquid fertilizer will ensure several repeat bursts of flowering through until autumn. Prune hard after flowering to avoid a sprawling, top-heavy effect.

## *Cheiranthus*

WALLFLOWER

***Brassicaceae*** (Mustard family).

***Scent:*** Sweet, spicy, similar to Stock.

***Description:*** This group of perennial or biennial herbs has been used for centuries as a cottage garden plant. A wild flower in England and Europe, its fragrant gold and crimson flowers are a pretty feature in old limestone quarries or by tumbledown stone walls. It belongs to the same family as Stock and Sweet Rocket and John Gerard actually referred to wallflowers as 'Yellow Stock-

▶ The rambling beauty of a farm garden.

gillyflowres' because of the similarity in their scent. They were well-loved by seventeenth century gardeners, being also known as 'chevisuance', an old English word meaning 'comforter'. Parkinson described them prettily, saying: ' . . . the sweetness of the flowers causeth them to be used in nosegays and to deck up houses'. The fact that wallflowers were so commonly used in posies gave rise to their botanical name, *Cheiranthus*, or 'hand-flower'.

***Species:*** Most wallflowers share the same rich, spicy scent, but the colors available vary widely. They are ideal for interplanting with other annuals and bulbs, especially daffodils. Choose from *Cheiranthus cheiri* (yellow, orange and golden-brown flowers; 2ft/60cm) or one of the dainty miniature varieties such as *Erysimum hieraciifolium* or Siberian Wallflower (deep orange and gold flowers; 15in/40cm). When selecting a variety, look for a strain that is double-flowered because this will be even more fragrant. The United States derived strains 'Fair Lady' and 'Early Wonder' are worth seeking out, as they provide double flowers in a variety of soft, silvery pastels.

***Cultivation:*** Sow wallflowers from seed in late spring each year, then set seedlings about 6–9in/15–20cm apart in sheltered spots throughout the garden in late summer to early autumn. Plants will then be ready to flower in the following summer. To encourage bushy growth, pinch out tips in late summer. Wallflowers will, not surprisingly, flourish against a warm stone wall; they also like sunny flower beds and the dwarf varieties are a pretty sight in tubs or window boxes. They seem to prefer slightly chalky soil.

## Choisya

MEXICAN ORANGE BLOSSOM

*Orange Flower Water*—Take one handful orange flowers and put them to about a quart of water and a quarter of a pound of sugar, let this stand and steep about half an hour, then take your water and flowers and pour them out of one vessel into another till such time as the water hath taken the scent and tast of the flowers, then set it in a cool place a cooling and you will find it a most excellent scented water.

*A Perfect School of Instructions for the Officers of the Month*, by Giles Rose, one of the Master Cooks to Charles II, 1682

***Rutaceae*** (Rue family).

***Scent:*** Heavy, sweet, orange-like.

***Description:*** These small evergreen trees are popular garden shrubs in most warm climates, as well as in their native Mexico. The glossy, dark green leaves are ornamental as well as aromatic, and emit an astringent, orange-like bouquet when crushed. The flowers are small, white and sweetly-scented and are clustered near the ends of branches, causing a slightly drooping appearance when the shrub is weighed down with its delicious-smelling burden.

***Species:*** *Choisya ternata* (white flowers; 5–10ft/1.5–3m) is the key species.

***Cultivation:*** Choisya will take some shade provided the area is warm. It must be sheltered from hard frost, but will tolerate a cold winter if confined to a garden. Choisya is an ideal choice for a greenhouse or as a tub plant in a tiled patio or courtyard. It is a quick growing plant and should flower during the first season it is planted. Cuttings of the young wood may be readily taken and struck in good quality potting mix before being established in a garden.

## Cistus

ROCK ROSE

*Dry roses put to the nose do comfort the brayne and the herte and quickeneth the spyryte.*

R. Banckes, *A Little Herball*, 1525

***Cistaceae*** (Rock Rose family).

***Scent:*** Sweet, balsamic.

***Description:*** A beautiful group of shrubs, particularly popular in northern gardens because they are evergreen and the scent of the leaves will lift the spirits in a winter garden. In southern Europe, the resinous gum secreted by the leaves is harvested and distilled to form labdanum, a key ingredient for many perfumes and medicines. A decoction of the roots was once used by Arabic doctors to staunch bleeding. Rock roses have also enjoyed popularity as ornamentals in the fragrant garden; most of the species' flowers are extremely beautiful single blooms and last well when cut. Rock roses grow extensively on Corsica and it is said that Napoleon, returning to his native country, would close his eyes and sniff the warm sea breeze to determine how close the boat was to shore.

*Cistus purpureus*

***Species:*** The tallest variety is *Cistus* x *aguilari* (white/gold flowers; 8ft/2.4m). The Gum Cistus, *C. ladanifer* (white/maroon flowers; 3ft/1m) is a major source of labdanum, releasing fragrance from the sticky leaves throughout summer and autumn. Other varieties include *C.* x *cyprius* (white/scarlet flowers; 6ft/1.8m); *C. laurifolius* (white/pale green flowers; 6ft/1.8m), which has attractive laurel-like leaves; *C. creticus* (mauve/purple flowers) and the delightful *C.* x *skanbergii* (bright pink flowers), the foliage of which has a spicy, cedar-like scent when bruised.

***Cultivation:*** Plants may be propagated by taking cuttings from an established shrub in late summer, then striking them in a mix of sandy loam before transferring them to an ornamental pot or into the garden. Rock roses prefer sandy soil and do not like to be disturbed. An ideal position is in a coastal garden with a sheltered aspect.

## *Citrus*

ORANGE, LEMON, LIME, TANGERINE, GRAPEFRUIT AND MANDARIN

TO MAKE SWEET WATER

*Take Damaske Roses at discretion, Basil, Sweet Marjoram, Lavender, Walnut Leafs, of each two handfuls, Rosemary one handful, a little Balm, Cloves, Cinnamon, one ounce, Bay leaf, Rosemary tops, Limon and Orange Pills of each a few; pour upon these as much white wine as will conveniently wet them and let them infuse ten or twelve days; then distil it off.*

*Receipts in Physick and Chirurgery,*
Sir Kenelm Digby, 1668

*Citrus sinensis*

***Rutaceae*** (Rue family).

***Scent:*** Extremely rich, spicy and sweet variations of the predominant fragrance, neroli.

***Description:*** A group of small trees and large shrubs with extremely fragrant white flowers and aromatic leaves. The fruit also has a distinctive smell, not only in the flesh and juice, but most noticeably in the rind, due to tiny pits or glands full of essential oil. Trees often sport fruit, flowers and buds simultaneously throughout the year for the buds take up to a year to ripen. Native to China and Japan, citrus trees were introduced to Mediterranean areas from Palestine and Spain in the eleventh century. Saint Dominic was said to have planted the first orange tree in Rome in 1200. Louis XIV of France was particularly fond of the rich, spicy scent of citrus trees and, accordingly, ordered an orangery filled with the potted trees to be constructed at the Palace of Versailles. One of these trees had been grown from seed collected by the Queen of Navarre in 1421 and survived until the late nineteenth century, making it one of the oldest known trees in the world.

The genus is hardy and will grow in most conditions. In northern climates, however, they should be kept in a greenhouse for protection during winter. John Evelyn, in his *Kalendarium Hortense* (1666) concurred with this strategy for successful citrus growing: 'Never expose your oranges, limons and like tender trees, whatever seasons flatter, 'til the Mulberry puts forth its leafe, then bring them boldly out of the greenhouse; but for a fortnight let them stand in the shade of a hedge.'

*Citrus aurantium*, the Seville, or bitter, orange, is cultivated in Spain. This is the variety used for commercial marmalade production and also for the extraction of neroli oil, an ingredient of Eau de Cologne. Neroli oil was named in honor of the sixteenth century Princess of Nerola who used it extravagantly to scent her clothes and rooms. The oil obtained from the bergamot orange is also used in perfumery and in the production of Eau de Cologne.

***Species:*** Different citrus varieties will contribute beauty and fruit as well as delicious fragrance to a garden. Choose from *Citrus limon* 'Meyer' (white/pink tinted flowers; 2ft/60cm) or Meyer Lemon, which is an ideal tub plant; *C. aurantium* var. *melitensis* (white flowers; 4–5ft/1.2–1.5m), the Portuguese Orange, with its red-skinned fruit; *C. grandis* (white flowers; 6ft/1.8m) or Shaddock Tree; *C. aurantium* (white flowers; 6–10ft/1.8–3m) or Seville Orange; or *C. limon* (white/pink flowers; 6ft/1.8m), the Common Lemon. Other species worth seeking out are the Chinese Tangerine or 'Satsuma Orange' *C. nobilis unshiu* with its bright orange, perfectly round fruit; the Sweet Orange, *C. sinensis* and the Jamaican Lime.

***Cultivation:*** Citrus trees require temperate to subtropical conditions.

NOTE: oranges tolerate humidity better than lemons, so are a better choice for a greenhouse. Citrus trees do not like frost at all. An acid soil is essential so feed with a commercial formula or water with a mild vinegar solution. Citrus are subject to many and varied pests, particularly scale insects, so regular checks on fruit and foliage are necessary.

### LIME FLOWER AND ELDER BLOSSOM MASSAGE OIL

❖

*1 handful each dried lime flowers and elder blossoms*
*7fl oz/200ml almond oil*
*3½fl oz/100ml apple cider vinegar*

*Pack flowers into a glass jar and cover with oil. Leave in a cool, dark place for 3 weeks, shaking occasionally. Heat the mixture, strain out flowers and add the vinegar. Pour into a sterile jar and cool before capping.*

▸ A Crab Apple Tree edged by miniature Box.

## *Clematis*

*Coming to kiss her lips, (such grace I found)*
*Meseemed I smelt a garden of sweet flowers:*
*That dainty odours from them throw around*
*For damzels fit to deck their lovers bowers.*

Edmund Spenser, sixteenth century

***Ranunculaceae*** (Buttercup family).

***Scent:*** Soft and sweet; especially pronounced at night.

***Description:*** Of all the scented plants, this group of summer-flowering shrubby climbers is a favorite throughout the world. Noted for the profusion of their star-like flowers and, in some areas, rapid growth, the different species of clematis offer a variety of sweet fragrances—from vanilla and almond types to a soft, chocolate-like scent. The word clematis derives from the Greek *klema,* meaning 'vine branch', referring to the way clematis climb using their curled leaf stalks like a vine's tendrils. Clematis is native to the British Isles and the variety *Clematis vitalba* was once a very common sight in English hedgerows. John Gerard wrote that it made 'a goodly shadow' beneath which passers-by could rest and, accordingly, named it 'Travelers' Joy'. With its firmly clasping habit, clematis became an emblem of romantic devotion and Robert Herrick likened it to his mistress:

*You are a full-spread faire-set vine,*
*And can with tendrils, love intwine,*
*Yet dry'd, ere you distil your wine.*

In his *Paradisi in Sole Paradisus Terrestris* (1629), John Parkinson dubbed clematis Virgin's Bower, because it was popularly used to grow over arbors where young girls might sit and be protected from harsh sunlight or the stares of passers-by. He added that 'country women just call it "Love" from its habit of tightly embracing other plants against which it grows.'

***Species:*** Not all clematis species are perfumed, so for best results, select from the following:

*Clematis paniculata*

*Clematis afoliata* (yellow flowers with strong, daphne-like scent); *C. cirrhosa* (white or greenish-yellow bell-shaped flowers with honeyed scent; 8ft/2.4m), which Gertrude Jekyll described having seen in Algiers where 'groves of Prickly Pear . . . [were] . . . wreathed and festooned with the graceful tufts of bell-shaped flowers and polished leaves of *Clematis cirrhosa*'. Also lovely are *C. flammula* (white/cream flowers with scent of meadowsweet) and *C. rehderiana* (pale yellow flowers with cowslip-type perfume).

Most clematis are summer-flowering, however, *C. cirrhosa balearica* is a pretty choice for winter because it flowers just before Christmas in the Northern Hemisphere. Autumn-flowering *C. paniculata* with its creamy vanilla-scented flowers is another good choice. *C. montana* is the best known and one of the most common species because it is prolific, adaptable and energetic, grasping its way to 30ft/10m or more in a few years. There are also several very striking hybrids.

***Cultivation:*** Clematis enjoy cool root runs, so protect the base of the plant by setting lavender or other climbing plants close by to give shade to the roots. Clematis like some lime rubble in the soil and should also be fed once a year with well-rotted manure. Train clematis over a pergola or allow it to ramble over an old tree trunk or up a wall. Since flower buds only appear on year-old wood, do not prune the first year. Subsequently, trim older wood enough to avoid tangling each spring, then prune hard after flowering. Clematis is reasonably root-hardy, but the branches will not tolerate excessive frost. Plants are readily propagated from cuttings and should remain in pots for at least one season before being transplanted to the garden.

### SCENTED SORBET

A refreshing summertime dessert—to create a splendid centerpiece for a party, serve sorbet in an ice mold. Simply arrange flowers and petals in a ring mold, add water and freeze. Decant onto a serving platter and heap scoops of sorbet into the center.

❖

*5½oz/175g sugar*
*7fl oz/200ml water*
*7fl oz/200ml champagne*
*4 tbspns rose petals, lavender flowers or violet blooms*
*juice of 1 lemon*
*2 large (2fl oz/60g) egg whites, whisked*

*Dissolve the sugar in the water and champagne. Bring to the boil and infuse rose petals or other flowers for 15–30 minutes. Add lemon juice, strain the mixture, and pour it into a metal freezer tray. Chill until the mixture begins to freeze, then fold through egg white(s). Freeze until firm.*

❖ ❖ ❖

## *Convallaria*

LILY OF THE VALLEY

### TO MAKE SPIRIT OF LILLEY OF THE VALLEY

*Gather your Lilley-of-the-Valley Flowers, when they are dry, and pick them from the Stalks; then put a quarter of a pint of them into a Quart of Brandy, and so in proportion, to infuse six or eight days; then distil it in a cold Still, marking the Bottles, as they are drawn off, which is first, second and third, etc. When you have distill'd them, take the first, and so on to the third or fourth, and mix them together, till you have as strong as you desire; and then bottle them, and cork them well, putting a lump of Loaf-Sugar into each Bottle.*

*This serves in the room of Orange-Flower Water in Puddings, and to perfume Cakes; though it is drank as a Dram in Norway.*

R. Bradley, *The Country Housewife and Lady's Director* (1732)

***Liliaceae*** (Lily family).

***Scent:*** Exquisitely sweet and slightly spicy.

***Description:*** This genus comprises a single species of bulb, which is commonly known as lily of the valley. Native to the British Isles, northern Europe and the northeastern mountains of America, the exquisite fragrance of the small, bell-shaped flowers should ensure this plant's place in any scented garden. In his *New Herbal* (1578), Henry Lyte called it '. . . the Lily Convall with flowers as white as snow and of a pleasant strong savour. The water of the flowers comforteth the hearte . . . and doth strengthen the memorie.' Other medieval herbalists claimed a tincture of the flowers would regulate a disturbed heartbeat and, quite recently, this remedy was used for treating British soldiers in the First World War who had been subjected to nerve gas. John Lawrence, in *The Flower Garden* (1726) said: 'The Convall-Lily is esteemed to have, of all others, the sweetest and most agreeable perfume; not offensive nor over-bearing, even to those who are made uneasy with the perfumes of other sweet scented flowers.'

*Convallaria* sp.

***Species:*** *Convallaria majalis* is a perennial with strong, twined rootstock, tough enough to push through even gravel. Once established in a garden, lily of the valley should increase steadily, due to the sturdy nature of its creeping rootstock. The majority of lily of the valley plants have bright green lance-like leaves, but there is another form, 'Variegata', with striking lengthwise golden stripes. (NOTE: although lily of the valley generally prefers a shady patch, the 'Variegata' form needs a little more sun in order to retain its foliage color.)

If you are as enamored of the dainty flowers as Nicholas Culpeper was—he described them as being '. . . like little bells with turned up edges'—then look for the double form, *C. majalis* 'Flore Pleno', which bears up to twenty buds per stem.

***Cultivation:*** Lily of the valley is originally a woodland plant and if you remember this, you are unlikely to err in growing it. The plants prefer dappled light and make ideal underplantings for trees and shrubs or around doorways, arches and gates. Loose, rich soil with plenty of humus worked in will give the best results. After flowering is finished each year, give plants a generous helping of mulch and/or compost.

### LILY OF THE VALLEY SCENTED BEADS

The fragrance of scented jewelery intensifies with the warmth of the skin and should last many years.

❖

*4 tbspns dried lily of the valley flowers*
*4 tbspns orris root*
*4 tbspns gum tragacanth*
*hat pin and fishing line*
*lily of the valley essential oil*

*Blend the flowers and orris root, add the gum and mix to a paste. Rub your hands with the oil, pick up pinches of the mixture and roll in your palms to form beads. Place on waxed paper and leave overnight. Dip a hat pin in oil. Pierce each bead and thread onto fishing line. Wrap in tissue and store in dark, dry place for 6 weeks, or until hardened.*

## *Coriandrum*

CORIANDER

CORIANDER WATER

*Take a handful of Coriander seeds, break them and put them into about a quart of water and so let it stand, put in a quarter of a pound of sugar and when your sugar is melted and the water well taken the taste of the seeds, then strain it out through a cloath and drink it at your pleasure. You may do the same with aniseeds.*

*A Perfect School of Instruction for the Officers of the Month*, by Giles Rose, one of the Master Cooks to Charles II (1682)

***Apiaceae*** (Parsley family).

***Scent:*** Highly aromatic, pungent, spicy.

***Description:*** This genus comprises a single species of annual herb, *Coriandrum sativum*. The glossy dark green leaves and small mauve flowers are attractive and have many culinary applications. The fresh leaves are much used as a garnish in China and the Middle East. However, it is the seeds that have been the most valued for adding flavor and aroma to food, especially pickles, chutneys and curries. Their taste is a cross between ginger and lemon, with a hint of almost floral sweetness. The ground and powdered seeds are also used in the commercial production of some brands of marmalade and of gin; it also adds a warm spicy aroma to biscuits and pastry. These seeds, which have a most unpleasant smell before ripening, were ordained as one of the Bible's 'bitter herbs', to be eaten during the Passover. They were more often prepared as sweetmeats, though, and in Exodus (xvi:31) we learn that '. . . the house of Israel called the name thereof Manna: and it was like Coriander seed, white; and the taste of it was like wafers made with honey.' Coriander was introduced to England from the Levant by the Romans, where it was widely grown during medieval times. 'Mayster Jon Gardener', in his *The Feate of Gardenynge* (1440) lists it as commonly grown in herb gardens. It was once grown commercially in Kent and Essex in Britain, as well as in America and India, but the best seed is popularly thought to come from North Africa and Egypt.

Coriander featured as a medicinal herb, an infusion being recommended for indigestion and worms and in cosmetic preparations for softening and bleaching the skin. In Wylliam Turner's *A newe Herball* (1551), we find reference to coriander's use in treating erysipelas, a particularly virulent skin disease: 'Coriandre layd to wyth breade or barly mele is good for saynt Antonyes fyre.' The seeds were also an ingredient in the famous 'Carmelite Water', a toilet preparation concocted by the Parisian sect of Carmelite monks in the early seventeenth century.

Coriander flowers

***Species:*** There is a single species, *Coriandrum sativum* (pale pink/mauve flowers; 3ft/1m).

***Cultivation:*** Coriander is a hardy annual. It prefers hot sun, so a bright warm corner of the herb garden is best. The seeds may be sown directly into the garden in lines about 8in/20cm apart in early spring provided the soil is light and well-drained. Seeds are sometimes slow to germinate, however, so to ensure successive cropping, treat coriander as a biennial and plant in autumn as well. Herbalists from times past felt that 'Coriander, Mallowes, Chervil and Dill love to grow together' and indeed they do seem to flourish near one another. Harvest the seeds in late summer, scattering a few as you do so and store in airtight containers.

### SPICY CORIANDER SPREAD

An unusual and flavorsome dip or spread. Serve with chunks of crusty, wholemeal, homemade bread, piled into a wicker basket.

❖

*4 tbspns chopped fresh coriander (cilantro)*
*1 tspn ground cumin*
*1 clove garlic, peeled and crushed*
*2 tspns paprika*
*½ tspn cayenne pepper*
*juice of ½ lemon*
*1 small Spanish (purple) onion, minced*
*1 tbspn tomato purée (sauce)*
*7oz/200g cottage or ricotta cheese*

*Blend all the ingredients in a food processor on low speed until smooth. Spoon into an earthenware pot and chill before serving.*

❖ ❖ ❖

## *Cornus*

DOGWOOD

***Cornaceae*** (Dogwood family).

***Scent:*** Soft and sweet.

***Description:*** A group of deciduous small trees or shrubs, native to northern Europe, Asia and America. Dogwood is a most attractive plant and contributes to the visual impact of a garden with striking reddish-purple stems and creamy white flowers or pink flowers. These have a delicate and sweet perfume that attracts bees and butterflies to the garden. The olive-shaped 'heps' that follow flowering may be made into a fragrant jelly. Much folklore attends the pretty dogwood: in East Prussia if dogwood sap is dripped onto a handkerchief and placed under a pillow, then the sleeper's dreams will be fulfilled. An old Newfoundland superstition among seafaring folk runs: 'I'd as lief cut off my right hand as cut down a maiden dogberry tree—a man is sure to drown if he does so.'

***Species:*** There are several species well suited to cool climate gardens, which are all highly decorative as well as fragrant. *Cornus macrophylla* or the Large-Leaf Dogwood is very colorful. *C. florida* and *C. nuttali* are also very sweet-scented. Even though dogwood is a deciduous plant, it is worth placing it where it may also be enjoyed during winter, for the crimson stalks make a most attractive foil for any nearby evergreen foliage.

***Cultivation:*** A cool climate is required to successfully grow dogwood. Plants should be hard pruned after fruiting each year to encourage new growth the following season. Dogwood likes to have its roots well protected from excess heat in summer, so underplant with bulbs, winter-flowering pansies or violets. Mixing a little grit through the soil will ensure good drainage; dogwood's roots do not like being waterlogged.

## *Crinum*

DARLING LILY, MURRAY LILY, SWAMP LILY

***Amaryllidaceae*** (Amaryllis family).

***Scent:*** Rich, soft, honeyed.

***Description:*** A group of plants native to Australia, parts of the American South and Southeast Asia, crinums are distinguished by their large-sized, long-necked bulbs. In Natal, they were once used as a powerful emetic. They are intensely fragrant and their scent intensifies as evening falls.

▸ A stone garden seat nestles beneath the trunk of a Manchurian Flowering Pear.

*Crinum moorei*

Crinums are spectacular and elegant assets to a scented garden once they are established, sporting bright green, strap-like leaves up to 3ft/1m with long and large, graceful funnel-shaped flowers. Crinums are a good choice for planting under a verandah or window that is well-placed for evening entertaining as the flowers, which droop downwards bashfully during the day, actually 'flare' upwards in the evening, releasing scent in bursts and attracting the night-flying *Lepidoptera* moth for pollination.

***Species:*** *Crinum pedunculatum,* or the Swamp Lily (white flowers; 2ft/60cm) is native to Australia, being particularly well known along the banks of the Darling and Murray Rivers. *C.* x *powellii* (rosy pink flowers; 2ft/60cm) is the most hardy and, therefore, most suitable for English or North American gardens. *C. bulbispurmum* (pale and dark pink flowers; 2ft/60cm) and *C. asiaticum* (white and pink flowers; 2ft/60cm) are more suited to warmer parts of the world. *C. moorei* (pink flowers; 2ft/60cm) is a summer-flowering bulb. *C. kunthianum* (white/crimson flowers; 3ft/1m) is found throughout the American South where it is well loved for its lushly spicy aroma.

***Cultivation:*** Crinums require a well sheltered spot in the garden. Plant the bulbs 12in/30cm apart in late spring to midsummer in a rich, well-drained soil so their necks are just at ground level. This will allow them to settle in over the winter period; ensure they are very well mulched if the area is subject to frosts.

## SCENTED COAT-HANGERS

❖

*1oz/25g orange blossom*
*1oz/25g lemon verbena*
*½oz/15g lemon balm*
*½oz/15g lemon thyme*
*2 tbspns cinnamon, crushed*
*1oz/25g orange zest, dried and crushed*
*2 tbspns orris root powder*
*2–3 drops orange oil*

*Mix all the dry ingredients together in a china bowl and stir through the oil. Fill a muslin bag the length of the coat-hanger with the mixture, then sew it in place on top of the hanger. Cover with pretty silk, organza or sprigged muslin. A finishing touch—swing a tiny 'swete bagge' of lace-trimmed matching fabric from the hook.*

*Crocus* sp.

## Crocus

*Pare saffron plot, forget it not*
*His dwelling made trim, look shortly for him*
*When harvest is gone, then saffron comes on;*
*A little of ground brings saffron a pound*

Thomas Tusser, *Five Hundred Points of Good Husbandry*, 1580

***Iridaceae*** (Iris family).

***Scent:*** Sweet, honeyed; some species have mossy or musky overtones.

***Description:*** A group of 75 bulb species, native to southern Europe. Crocus, especially the saffron crocus (*Crocus sativa*) are among the oldest of plants in cultivation and much folklore surrounds them. For example, it was once thought that crocus could inspire love; an Austrian superstition, however, claims it is bad luck to pick the flowers as they sap a person's strength.

The crocus is believed to be the *Karkan* mentioned in the Bible's *Song of Solomon*. In ancient Rome, it was known as *krokas* and a promontory in what is now part of Turkey was dubbed *Korykos* to honor the saffron crocus, which was farmed there. The Romans used crocus to deck their temples and to flavor and color their food. The dried powdered petals were used to dye hair and clothing golden. Sweet-scented saffron water was sprinkled in the amphitheaters to counter the stench of blood. A favorite confection was saffron balls, where the powder was mixed with honey or rosewater and then rolled in chopped candied angelica. Saffron was also used with varying effect as a tonic and/or a medicine. Francis Bacon had a very high opinion of saffron and believed 'what made the English people sprightley was the liberal use of saffron in their broths and sweetmeats' although Turner warned darkly that: 'It is sterke poyson and will strongell a man and kill him in the space of one day.' Nowadays it is only used in very small quantities as a flavoring agent for cakes and Cornish pasties.

One very old legend has it that saffron crocus was introduced to Britain by a pilgrim who returned from the Holy Land carrying a bulb in his hollow staff. It certainly became a widespread and popular plant in England. In fact, so much of it was grown in Essex that a town was named in its honor—Saffron Walden. For many centuries, when kings and important personages visited this town, it was customary to present them with a fine silver tureen filled with saffron.

***Species:*** Different varieties of crocus flower through summer and winter. It is possible to plant different kinds throughout a garden and therefore enjoy the lovely scent throughout the year. *Crocus chrysanthus* (golden yellow, mauve or blue flowers, some with stripes or a feathery pattern) flower towards the end of winter and continue into early spring. Snow Bunting is a very pretty variety of *C. chrysanthus* with palest blue and white petals. *C. laevigatus fontenayi* (purple and white flowers) blooms right through winter. *C. imperati* (violet flowers with scarlet anthers) is a stunning choice for a spring flowering with its large flowers and honeysuckle-like scent. Roman women used the dried petals around the home to perfume linen and as stuffing for pillows and mattresses. *C. suaveolens* (white, starry flowers), or the Roman Crocus, is also a spring bloomer as is the delightful *C. angustifolius*, or Cloth of Gold Crocus. *C. sativus*, the famed Saffron Crocus, was described by John Parkinson in his *Paradisi in Sole Terrestris* (1629) as '. . . the true Saffron that is used in meates and medicines . . . the flowers are of a murrey or reddish-purple colour, having a shew of blue in them'. They are, indeed, a very beautiful inclusion to a scented garden, with their richly colored flowers appearing in mid-autumn.

***Cultivation:*** Crocus species will flourish in most climates, provided the soil is well drained and rich and, if subject to frost, well-mulched throughout the winter. A good idea for winter-flowering species is to plant them in a dense mat of green ground cover, such as phlox, so the roots are well protected. Crocus are an ideal potted plant and have long been grown outside for indoor use. In nineteenth century France the bulbs were placed in special globe-shaped forcing jars, filled with warm water. These were a fashionable accessory in many reception rooms where the warmth would help release the plant's fragrance. Some of the smaller species, such as winter-blooming *C. laevigatus*, are a charming sight in window boxes.

*Crocus* sp.

TO MAKE SYROP OF SAFFRON

*Take a pint of the best canary, as much balm-water and two ounces of English saffron; open and pull the saffron very well, and put it into the liquor to infuse, let it stand alone uncover'd (so as to be hot but not to boil) twelve hours, then strain it out as hot as you can and add to it two pounds of double refined sugar; boil it till it is well incorporated, and when it is cold bottle it; and take one spoonful in a little sack or small cordial, as occasion serves.*

E. Smith, *The Compleat Housewife*, 1736

## *Cyclamen*

*Cyclamen ought to be grown in every house, if it be true that wherever this plant grows no noxious spells can have effect. This plant is also called an amulet.*

Pliny the Elder

***Primulaceae*** (Primrose family).

***Scent:*** Pretty and sweet; variously like lily of the valley or vanilla, depending on species.

***Description:*** A group of plants with corm-like tubers, which are native to the British Isles, most European countries and North Africa. Most familiar of all is the large-flowered Persian Cyclamen, or *Cyclamen persicum,* which is the florists' favorite, even though it is not fragrant. The bouquet of the smaller, 'wild' varieties makes them the ideal choice for planting beneath deciduous trees or allowing to naturalize in the lawn of a scented garden. Country folk once regarded the heart-shaped, plum colored flowers of cyclamen, especially *C. purpurascens,* as a potent aphrodisiac. A decoction of the roots was used to relieve the pains of childbirth and an ointment, also made from the roots, was used as a vermifuge, i.e. to expel worms from the body. Cyclamen was once known as Sowbread because the plants were fed to pigs.

***Species:*** *Cyclamen purpurascens* (purplish/red flowers in summer) has the most sweetly scented flowers, reminiscent of lily of the valley. It was a great favorite in Elizabethan 'knotte' gardens. *C. creticum* (white flowers in late summer/early autumn) offers a carnation-like scent. Other choices include *C. hederifolium* (pale pink flowers, most attractive marbled white leaves). *C. cilicum*'s honey/vanilla scent and pretty pink flowers will enliven a tired autumn garden as will the heavily scented crimson-spotted white flowers of *C. cyprium.*

***Cultivation:*** Cyclamens appreciate semi-shaded conditions and should have some protection in gardens. Moisture is required for flowering but soil must be well drained or corms will rot. Gerard tells us that fine specimens of *C. hederifolium* and *C. europum* grew on limestone foundations in Kent and Sussex and cyclamens will thrive if a little lime is added to the soil. The tubers should be planted just below ground level—merely press them into the ground —and barely covered with a shallow mulch. Plant in early summer for flowering the following year. After flowering, cyclamens should be generously top dressed, with a good helping of leaf mulch, peat and lime rubble. An established cyclamen corm will increase over a period of 50 years or so and each corm will generate up to 100 flowers a year, so it is well worth the effort.

*Cyclamen hederifolium* (syn *neopolitanum*)

### Romantic Fan

*Once upon a time, ladies' fans conveyed an entire language privy only to lovers. Have a romantic flutter on a hot day by decorating your straw fan with a tiny posy of dried flowers and a sachet of potpourri tied on with a velvet ribbon. The scent is very refreshing as the fan sways to and fro.*

## Cymbidium

***Orchidaceae*** (Orchid family).

***Scent:*** Sweet and delicate, long-lasting.

***Description:*** Though few orchid species are fragrant, some of the cymbidiums bear the most beautifully scented flowers of all. They originated mainly in the valleys of the mighty Yangtze River in China and in Japan, they may be cultivated in most warm areas. First recorded in manuscripts of the Sung Dynasty (960–1279), the cymbidium's delicate scent was described as 'the scent of kings'.

*Cymbidium* hybrid

***Species:*** *Cymbidium suave* (yellowish green flowers) is an Australian native with a strong scent; *C. eburneum* (pure white flowers) is also very fragrant and attractive. Best known is the spring-flowering cymbidium (green-yellow flower with bright chartreuse markings), which is also powerfully fragrant.

***Cultivation:*** As with other members of the orchid family, cymbidiums require a rich compost, good drainage, a constant, warm temperature and broken sunlight. The roots are large and fleshy and need plenty of room if they are to be potted. Plants may be propagated by dividing up established plants.

## Cymbopogon

Lemon Grass

*. . . the grass that lay smelt so deliciously,*
*the buzzled bee went wandering where the honey sweet could be*
*And passers bye among the level rows*
*Stoop'd down and whip't a bit beneath his nose . . .*

John Clare

***Graminaceae*** (Grass family).

***Scent:*** Lemony, with verbena or rose overtones.

***Description:*** This group of perennial scented grasses from India and Southeast Asia have been cultivated for their oil since very early times. According to legend, as Alexander the Great rode his elephant along the Egyptian border, he was intoxicated by the smell of *Cymbopogon nardus* given off when the grass was crushed by the elephant's thundering feet.

The Egyptians, Greeks and Romans all used lemon grass oil or otto as a cosmetic and perfume.

*Cymbobogon* sp.

It is often sold to adulterate otto of roses in perfumery production. It was also one of the ingredients of the holy anointing oil with which Aaron and his sons were consecrated. Today it can often be found in Indian shops or health food stores, rather confusingly disguised as 'oil of geranium'.

Lemon grass is hardy and, with its long thin leaves, it is an attractive feature in a garden. A softly waving clump of lemon grass interplanted with nerines or scentless irises can look extremely beautiful.

***Species:*** *Cymbopogon citratus,* Lemon Grass (5ft/1.5m); and *C. nardus* (syn. *Andropogon nardus*), which grows about 12in/30cm tall, is strongly aromatic and the source of citronella oil.

***Cultivation:*** In their countries of origin, the lemon grass species grow in barren land with a minimum of water. These conditions are best for the plant and help produce the sweetest-smelling oil. The leaves are easily harvested and, dried and crumbled, may be used in potpourri or sachets about the house.

## *Cytisus*

BROOM

TO PICKLE BROOM-BUDS AND PODS

*Make a strong pickle of White Wine, Vinegar and Salt able to bear an Egg. Stir very well till the Salt be quite dissolved, clearing off the Dregs and Scum. The next day pour it from the Bottom, and having rubbed the Buds dry, pot them up in a Pickle Glass, which should be frequently shaken till they sink under it, and keep it well stopt and covered. Thus you may pickle any other Buds.*

John Evelyn, *Acetaria,* 1699

***Fabaceae*** (Pea family).

***Scent:*** Fruity, pineapple-like.

***Description:*** The broom family are tall, bushy shrubs, densely covered in flowers during late spring and early summer. The flowers were once a heraldic emblem and symbolized humility. The order of the knights known as *l'ordre du Genest,* founded by St Louis of France, wore chains made of gilt broom flowers and enameled *fleurs des lis.* These knights formed the French royal bodyguard and, in later years, broom became incorporated in English heraldry as well. Country folk thought broom was a witches' flower and, consequently, it was accompanied by much superstition. For example, if green broom is picked before it has bloomed, it is said the mother or father in the household will die within the year. An old Sussex proverb echoes this belief:

*If you sweep the house with blossomed broom*
*in May*
*You are sure to sweep the head of the house away.*

Yellow cytisus

With its clusters of many flowers, broom was considered a fertility charm and sprigs of it were tied to the clothing of a bride and groom to ensure many children. Broom's close relative, gorse, also had romantic connotations and it used to be said that: 'When gorse is out of bloom, then kissing's out of season.'

The tips of broom were used in herbal medicine, an infusion being a cure for dropsy. John Gerard, in his *Herball* (1597) said: 'That worthy Prince of famous memory, Henry VIII of England, was wont to drink the distilled water of Broom-flowers against surfeits and diseases thereof arising.' Unopened broom buds were pickled in brine and used in Tudor 'sallets' in much the same way as capers are today.

***Species:*** *Cytisus battandieri,* or Morocco Broom (15ft/4.5m) has bright golden yellow flowers with the delicious scent of pineapple and is very quick growing. *C. nigricans,* the oddly named Black Broom, also has striking yellow flowers, which are prettily accented with silvery green foliage.

***Cultivation:*** Cytisus requires full sun and protection from cold winds. Under a window sill or against a sun-warmed wall is ideal. After flowering, the branches should be hard-pruned to ensure a full crop of flower buds the next year.

## *Daphne*

***Thymelaeaceae*** (Mezereum family).

***Scent:*** Sweet, spicy, clove or violet-like, depending on species.

***Description:*** These dainty deciduous shrubs are among the most powerfully scented plants in the world. Most are native to the cool mountainous regions of Europe, though some varieties originated in China and North Africa. Daphne was named after the pretty nymph who was so desperate to escape Apollo's eager embraces that she begged to be turned to stone. Instead, the gods turned her into the scented plant. Like many scented plants, the flowers of most daphne species are pale so that night-feeding moths can find them in the dark and facilitate pollination.

***Species:*** Most people are aware of the popular *Daphne* x *burkwoodii* (soft pink flowers in starlike clusters; 3–4ft/1–1.2m); just a few flowers from this bush will richly scent a garden or room. However, there are many varieties of daphne and in my opinion a 'scented gardener' can experiment with several. Start with *D. odora,* or Chinese Daphne (mauve flowers with green leaves edged in white; 5–6ft/1.5–1.8m). Legend has it that a Chinese monk named this variety 'Sleeping Scent' when its spicy aroma awoke him from slumber. Other varieties to choose from include: spring-flowering British native *D. mezereum* (pink/purple flowers with orange stamens; 3–4ft/1–1.2m), which Gerard described as the dainty 'mezereon' and was widely grown in sixteenth-century gardens; the dwarf variety *D. cneorum* (rosy pink flowers, a semi-prostrate form), which is also known as the Garland Daphne, is lovely in a rock garden. More unusual daphnes, which can be most striking, are *D. laureola* (greenish white flowers; 4ft/1.2m),

▸ A place to reflect while enjoying your garden vista.

*Daphne odora*

*Datura suaveolens*

which has large flat flower heads and poisonous glossy black berries and *D. mezereum* (wine colored flowers; 15in/40cm) with orange red berries. Also lovely is the trailing prostrate form *D. blagayana* (white flowers; 15in/40cm), which may be trained as a 'mat' under a tree situated in lime-rich soil.

***Cultivation:*** Daphne has a reputation for being temperamental but this is mainly due to the fact that many species are slow-growing. A common problem is over-watering—daphnes must have good drainage. They prefer a sheltered position with partial shade and most species also enjoy a generous proportion of lime worked into their soil. It is possible to plant daphnes for enjoyment year-round: *Daphne odora* and *D. odora* var. *marginata* flower in early and late winter; *D. mezereum* blooms in spring and *D. cneorum* in summer. Daphnes resent pruning, nor do they like being transplanted or having their roots disturbed in any way.

## *Datura*

ANGEL'S TRUMPETS, DEVIL'S TRUMPETS OR WEDDING BELLS

*The giant Datura bares her breast of fragrant scent, a virgin white*
*A pearl amidst the realms of night . . .*

Bishop Heber

***Solanaceae*** (Nightshade family).

***Scent:*** Soft, exotic, lily-like.

***Description:*** A group of shrubs native to South America. Although quite lovely, in Peru and Mexico they are actually regarded as pests because they spread so rapidly through grazing land. Elsewhere, however, they are much enjoyed as exotic, free-flowering scented plants of great beauty. The large, funnel-shaped drooping flowers are striking and glamorous and, if climate permits, a datura is a must for training up over a pergola or onto a balcony or patio so passers-by can lift their faces to the intoxicating perfume.

***Species:*** Different datura species provide a variety of flower colors. Best known is *D. suaveolens* (white flowers; 6–12ft/1.8–3.6m) or Angel's Trumpets, but there is also the yellow *D. chlorantha*, mauve *D. stramonium* or English Thorn Apple and orange-red *D. sanguinea*. Some words of warning when planting datura: first, the fruits that appear after flowering are poisonous as is the foliage; secondly, the flowers of some species wither and turn brown on the plant before they fall.

***Cultivation:*** Being subtropical plants, daturas require a warm position with plenty of water. They may be readily raised from seed. Sow seed in early spring and fertilize well. Daturas may be grown in northern climates as they will tolerate light frost, however, if the climate is severe it is best to grow the plants indoors or bring them inside in tubs for the winter. Depending on the effect desired, daturas may be trained up a wall or frame for a tree-like look, or pruned and pinched back into a bushy shrub.

## *Delphinium*

***Ranunculaceae*** (Buttercup family).

***Scent:*** Musky.

***Description:*** Immortalized in A. A. Milne's sweet poem about the relaxing effect of 'delphiniums blue and geraniums red' upon a certain dormouse, delphiniums comprise a group of annual or perennial herbs.

In a well planned garden, created with the purpose of attracting bees and/or butterflies, delphiniums should figure strongly. The musky fragrance of the key species, *Delphinium brunonianum*, was once thought to be the favorite food of the musk deer, thus neatly explaining the scent of the creature's exudations, but this was not actually the case. As well as being fragrant, delphinium flowers retain their delightful pale blue color when they are dried, which makes them especially effective in potpourri.

*Delphinium* hybrid

***Species:*** *Delphinium brunonianum* (large, pale blue flowers with black centre; 12in/30cm) is native to the Himalayan area. Somewhat rarer is *D. leroyi* (pale blue flowers; 2ft/60cm), which grows on Mount Kilimanjaro. This species is also available in many other parts of the world, being used widely in cross-breeding programs because its perfume is more intense than that of *D. brunonianum*.

***Cultivation:*** Delphiniums may be propagated from seed. Sow in spring in a well-drained soil for a summer showing of flowers; cut back dead material in late autumn and mulch well to protect plants against winter frosts to which they are very susceptible.

*Dianthus barbatus*

## *Dendrobium*

SPIDER ORCHID

***Orchidaceae*** (Orchid family).

***Scent:*** Vanilla, musk or violet-like, depending on species.

***Description:*** A group of orchids that have an intoxicating fragrance. Native to Australia, dendrobiums are epiphytic plants, that is, they grow on trees or rocks rather than with their roots in water or soil. In fact the name, dendrobium, is derived from the Greek words for 'tree-life'. Dendrobiums provide a striking feature in a garden, particularly if affixed to another scented tree, such as frangipani. They have tough, narrow, thick leaves and racemes of exquisite, creamy-yellow or rose-tinted flowers.

***Species:*** The most fragrant varieties are *Dendrobium tetragonum* (white flowers and vanilla scent; 1ft/0.3m); *D. aemulum* or Ironbark Orchid (white flowers); *D. moschatum* (yellow and/or pink flowers with a distinctive musky scent; 6ft/1.8m); *D. linguiforme* (white, thread-like flowers), which is commonly called the Tongued Orchid due to its appearance; and *D. nobile* (white, pink and purple flowers).

***Cultivation:*** Dendrobiums are drought and frost-tender, requiring a constant temperature around 70°F/21°C and plenty of moisture. A protected, shady position, either in the branches of a tree or on a warm rock, is an ideal growing position. They may be potted for a short period of time, using a very small amount of compost.

## *Dianthus*

PINKS, SWEET WILLIAMS AND CARNATIONS

### TO MAKE CLOVE-GILLYFLOWER WINE

*Take six gallons and a half of Spring Water, and twelve Pounds of Sugar, and when it boils skim it, putting in the White of eight Eggs, and a Pint of Cold Water, to make the Scum rise; let it boil for an Hour and a half, skimming it well; then pour it into an Earthen Vessel, with three Spoonfuls of Baum; then put it in a Bushel of Clove-Gillyflowers clip'd and beat, stir them well together, and the next Day put six Ounces of Syrup of Citron into it, the third Day put in three Lemons sliced, Peel and all, the fourth Day tun it up, stop it close for ten Days, then bottle it, and put a Piece of Sugar in each Bottle.*

Sarah Harrison, *The Housekeeper's Pocket Book*, 1739

***Caryophyllaceae*** (Pink family).

***Scent:*** Spicy, cinnamon or clove-like.

***Description:*** This group of summer flowers is a very ancient one. Theophrastus, the Athenian 'father of medicine', named it *di anthus,* meaning 'flower of Jove' and it was held in high esteem as a lucky plant. It was the main flower used in making coronets and garlands, which gave rise to its country name 'coronation' from which, in turn, the word 'carnation' comes.

Other traditional names were clove gillofloures, gilly-flowers and sops-in-wine, the latter being a reference to their use in flavoring food, especially vinegar, wine and jams, during Tudor times. In *The Good Housewife* (1692), Thomas Tryon wrote that 'our conserve of the carnation Gillyflower is exceeding cordiall eaten now and then.' They were also called Clove July Flowers due to their habit of flowering at this time. A popular garden flower in England, carnations and pinks also featured widely in medieval art, being used to symbolize the tears of the Virgin Mary or as an emblem to ward off death. The pink's name derived from the old Celtic word *pic* meaning 'peak of perfection'. In his *Amorati,* Edmund Spenser likened his young wife's pretty eyes to 'pinks, but newly spread'.

Shakespeare made frequent mention of pinks and it is thought that John Gerard dubbed Sweet Williams in his honor. William Lawson, in *A New Orchard and Garden* (1618) wrote: 'I may well call them the King of Flowers, except the Rose . . . I have them nine or ten, several colours and divers of them as big as Roses.' John Parkinson, in *Paradisi in Sole Paradisus Terrestris* was even more effusive, asking 'what shall I say to the Queen of delight and of flowers, carnations and gillyflowers, whose bravery [hardiness], variety and sweet smell joyned together, tyeth everyone's affection with equal earnestness both to like and to have them.' For many centuries, gardeners and horticulturalists have experimented with this group of flowers and in the eighteenth century the famous 'laced' pinks were developed, with the pattern and color of their petals being thought similar to the Paisley shawls after which they were named. Some of the varieties, like the carnation *Dianthus caryophyllus* 'Robin Thain', have had the serrated edges of their petals bred out and others have, quite horribly, been turned blue or green according to the whims of florists.

***Species:*** The main groups of *Dianthus* are carnations, pinks and Sweet Williams. Carnations are derived from *Dianthus caryophyllus* and are hardy, strongly scented varieties. Worth seeking out are 'Robin Thain' (white flowers with crimson stripes); 'Royal Clove' (rich pink flowers); 'Lavender Clove' (magenta flowers); 'Coral Clove' (apricot flowers) and 'Oakfield Clove' (deep red flowers). Pinks are mainly derived from *D. plumarius* and are so named for their fringed or 'plumed' petals. Pretty choices include the fringed 'Dusky' (soft pink flowers); 'Miss Corry' (scarlet flowers) and 'Winsome' (pink flowers with a scarlet center). Specialist nurseries may even be able to help source some of the old-fashioned varieties of pinks, like the sixteenth century 'Nonsuch' (pink and plummy petals) or the old Irish 'Black Prince' (white flowers with a striking black middle and pronounced spicy scent). Sweet Williams, or *D. barbatus,* are a very old flower. Gerard wrote that 'the flowers . . . are kept and maintained in gardens more to please the eye than either the nose or the belly . . . not used in meat or medicine but are esteemed for their beauty to deck up gardens,

*Dianthus barbatus*

the bosoms of the beautiful, garlands and crowns for pleasure.' Many different colored varieties of *D. barbatus* are available, from all shades of red and pink through to variegated types.

***Cultivation:*** Dianthus may be grown from seed or propagated via cuttings. Most prefer full sun, although a few, notably *D. arenarius*, are not quite as hardy and will do better in semi-shade. A well-drained soil containing plenty of lime is essential. Add a sprinkling of dolomite to the soil in late spring to ensure a healthy crop of flowers.

### CLOVE PINK AND RHUBARB PIE

❖

*2 tbspns clove pinks, finely chopped*
*3½oz/100g rhubarb, coarsely chopped*
*4 tbspns caster (superfine) sugar*
*grated zest of 1 orange, plus juice of half*
*1 oz/30g flaked almonds*
*pinch allspice*
*4 sheets short pastry*

*Preheat oven to 350°F/180°C. Simmer clove pinks with rhubarb, sugar, orange zest and juice, almonds and allspice until fruit has softened. Strain mixture to remove juice. Line a pie dish with pastry and spoon the mixture evenly over it. Cover with an extra pastry sheet and decorate with almonds. Bake in the oven for 20–25 minutes.*

## *Erica*

HEATHER

***Ericaceae*** (Heather family).

***Scent:*** Tangy, invigorating.

***Description:*** A group of evergreen bushy shrubs native to cool climate areas, notably Scotland, Ireland and the European Alps. Only a few varieties have sweet-scented flowers and it is the

*Erica* sp.

heather's foliage that exudes the refreshing, tangy aroma pervading the air around Scottish moorlands and Irish bogs. For many, the 'bonnie, bloomin' heather' is as Scottish as haggis. Many noble Scottish families incorporated heather in their family coat of arms and sprigs were exchanged as lovers' troth tokens. An old Scottish superstition has it that burning heather will cause rain to fall. On the Continent, scented briar pipes are made from the roots of heather.

***Species:*** Most heather varieties will flower in temperate zones as well as in cooler areas. Eleanour Sinclair Rohde described *Erica lusitanica* (pink and white sweet scented flowers; 5ft/1.5m) as '. . . the quintessence of the scent of wild open spaces in Springtime'. Good choices for spring and/or summer flowering are *E. denticulata* (white, tubular flowers) or the Sweet Heath; the Irish Heath *E. erigena* (pink, honey-scented flowers; 4ft/1.2m); *E. fragrans* (purple, honey-scented flowers in some cultivars; 12in/30cm) and *E. arborea* (white, sweetly perfumed flowers; 10ft/3m). One species, *E. bauera* (pale greyish flowers, very delicately scented), will flower throughout early winter.

***Cultivation:*** Most heather species are reasonably hardy, requiring a sandy acid soil and an open sunny aspect. They do not respond well to overfertilizing and particularly dislike animal manures. A light compost, leavened with sand and leaves is best. Heathers may be propagated by root cuttings after blooming. They should also be lightly pruned after flowering.

## *Eriostemon*

WAX FLOWER

***Rutaceae*** (Rue family).

***Scent:*** Fresh, slightly pungent.

***Description:*** A group of evergreen shrubs native to Australia with small white, starry flowers and thin dark green leaves. The flowers have a faint but very sweet scent in comparison with the aromatic leaves.

*Eriostemon australasius*

***Species:*** *Eriostemon myoporoides* (clusters of white flowers; 3-4ft/1-1.2m) or *E. buxifolius* (white flowers; 3ft/1m) are the main varieties. (*E. australasius* has a light scent on warm days.)

***Cultivation:*** Eriostemon tolerates most conditions, but is not frost-hardy. A well-drained soil is preferred and some root protection is also required so mulch plant surroundings well or set with a dense ground cover. In colder climates, eriostemons are an excellent choice for an indoor plant, as they will survive quite a long while if potted. If they are kept indoors, they require regular misting so the leaves will release their scent.

## *Escallonia*

***Saxifragaceae*** (Rockfoil family).

***Scent:*** Foliage is resinous and pungent; flowers have a sweet, faint perfume.

***Description:*** A group of evergreen shrubs and trees, often seen planted as a scented windbreak in seaside gardens. Being quite compact and densely bushy, escallonias are good screening or hedging plants and excellent for withstanding seaborne salt winds and spray. They are native to South America and their name was derived from that of the Spanish botanist, Professor Escallon, who first cataloged the species.

***Species:*** *Escallonia bifida* (white, honey-scented flowers; 12-15ft/4-5m) is a neat, compact shrub, which is widely available in nurseries. More exotic, and difficult to source is *E. macrantha,* which bears bright red flowers in summertime.

***Cultivation:*** Escallonias are an ideal hedging plant; they also like growing against walls. A sandy soil is preferred and regular hard-pruning is essential, otherwise they will become sprawling and ungainly.

## *Eucalyptus*

GUM TREES

***Myrtaceae*** (Myrtle family).

***Scent:*** Pungent, honey, peppermint or lemon-scented, or camphoraceous, depending on species.

***Description:*** A large group of trees and shrubs native to Australia and New Zealand. The genus name, *Eucalyptus,* refers to the appearance of the flowers, which emerge from a membraneous skin or 'cup' (the Greek *eucalyptos* meaning 'well-covered'). Eucalypts are among the most aromatic plants in the world and the volatile oil given off by the leaves has powerful healing, disinfectant and antiseptic properties, particularly when used as a douche or gargle. The fragrance of eucalyptus is head-clearing, sharp, and immensely refreshing.

*E. globulus* was dubbed the 'fever tree' because it grew in unhealthy, swampy areas and noticing this, Mrs Wilder wrote of a 'genus of immense, fast growing trees called Gum Trees because of the quantity of gum that exudes from their trunks. The thick leathery leaves give off balsamic odors supposed to increase the healthfulness of districts where they thrive'. Baron Ferdinand von Mueller, the controversial Director of the Botanical Gardens in Melbourne, Australia, was the first to suggest that eucalyptus oil had medicinal qualities. He sent seeds to Algiers where it was found that not only did the trees exude a refreshing and antiseptic aroma in marshy, illness-riddled districts, but the trees' roots helped dry out the unhealthily water-logged soil. In Sicily, eucalyptus trees were similarly planted as a malaria preventive. The medicinal qualities of eucalyptus oil were given the official seal of approval in the 1885 edition of the British *Pharmacopeia.*

***Species:*** *Eucalyptus globulus* (100–200ft/30–60m) is the best known variety. It is also the fastest growing tree in the world, averaging a height increase of 7–8ft/2–2.5m per year. More practical choices, however, do abound: *E. gunnii* (30ft/10m; decorative bluish leaves) or the hardy Tasmanian Gum; *E. viminalis* (30–200ft/10-60m), the Ribbon Gum so beloved of the Australian koala; the extremely fragrant Blue Gum *E. coccifera; E. torquata* (25ft/8m), which has delightful apricot flowers, and the lovely Lemon Scented Gum, *E. citriodora* (50ft/15m). Eucalypts grown for honey-making are *E. saligna* and *E. melliodora.*

*During the Second World War, wives and sweethearts would slip gum leaves in the parcels and letters they sent homesick Australian servicemen. When they opened their mail, Australians would cluster together and set a match to the leaves, inhaling the unmistakable aroma of the Australian bush.*

## *Eucharis*

AMAZON LILY

***Amaryllidaceae*** (Amaryllis family).

***Scent:*** Exotic, powerful, lily-like.

***Description:*** A group of tropical bulbous plants, native to the mountain forests of South America. Colombian natives revered the lovely cool white flowers and would present them to a girl on her betrothal. They believed the plant would give her strength to resist temptation and to give birth to vigorous children as a result of her chaste love for her husband. Similarly, the first lily found by a male member of the tribe was thought to bring strength and faith to its finder. The flowers are set in clusters and resemble the narcissus, being white with a tint of green and surrounded by dark green strappy leaves. They are powerfully-scented and long-lasting.

***Species:*** *Eucharis grandiflora* (large, 6in/15cm across, white flowers; 2ft/60cm) is a richly fragrant plant for a tropical garden or greenhouse. There is also a miniature form, 'Fosteri', which is available in some specialist nurseries or by mail order.

***Cultivation:*** Although eucharis varieties are really only suited to tropical gardens, they can be

▶ 'The rose looks fair, but fairer we it deem
For that sweet odour that doth in it live.'

included in a scented garden because they make an ideal plant for a sheltered position in a shadehouse or enclosed greenhouse. They can remain potted indefinitely as long as the container is large enough, for they have a complex root system and each bulb measures 3in/7.5cm alone. Terracotta or clay pots rather than plastic ones are recommended, as they 'breathe' better. Eucharis require plenty of moisture and steady warmth to prompt growth; a drying out period after flowering is also a good idea.

## *Foeniculum*

FENNEL

*A savoury odour blown,*
*Grateful to appetite, more pleased my sense*
*Than smell of sweetest Fennel . . .*

Milton, *Paradise Lost*

***Apiaceae*** (Parsley family).

***Scent:*** Pungent, hay-like.

***Description:*** A hardy perennial, fennel is a beautiful and useful plant in a scented garden. It has had many applications through history: as a restorative medicine, a strewing herb, a snake and flea repellent, a slimming aid, an ingredient in potpourri and bedding, and as a tea to increase the flow of milk in nursing mothers' breasts. Fennel is also a very beautiful and ornamental plant, with its bright green feathery leaves that turn bronze later in the season and its golden flowers, which appear in summer. The genus name, *foeniculum,* was bestowed by the Romans, in deference to its wholesome aroma (*foenum* meaning 'hay'). In medieval times, poor folk used fennel to make unsavory or rancid food palatable. In his *Theatricum Botanicum* (1640), John Parkinson writes that:

*The leaves, seede and rootes are both for meate and medicine; the Italians especially doe much delight in the use thereof, and therefore transplant and whiten it, to make it more tender to please the taste, which being sweete and somewhat hot helpeth to digest the crude qualitie of fish and other viscous meates. We use it to lay upon fish or to boyle it therewith and with divers other things, as also the seeds in bread and other things.*

*Foeniculum vulgare*

Fennel was also thought to be a medicinal herb and has long been renowned for its supposed ability to restore the eyesight. In *The Goblet of Life,* Longfellow refers to this aspect of the herb:

*Above the lower plants it towers,*
*the Fennel with its yellow flowers;*
*And in an earlier age than ours*
*Was gifted with the wondrous powers*
*Lost vision to restore.*

Fennel was also employed as a witchcraft preventive and hung over door lintels on Midsummer's Eve to ward off evil spirits. The very old *Nine Herbs Charm*, thought to be used by the Druids were:

*Thyme and Fennel, two exceeding mighty ones.*
*These herbs the wise Lord made.*
*Holy in the Heavens; He let them down,*
*Placed them, and sent them into the seven worlds*
*As a cure for all, the poor and the rich*
*It stands against pain, it dashes against venom,*
*It is strong against three and thirty,*
*Against the hand of an enemy and against the hand of the cursed,*
*And against the bewitching of my creatures.*

Fennel has a sweetish taste and is traditionally used to flavor cordials, pickles, vinegars and liqueurs, as well as salads, poultry and fish. Some of the oilier fish, such as mackerel and herring, are particularly well served by fennel as it does aid the digestion. In addition to its carminative properties, fennel acts as a diuretic and has long been favored by slimmers. Athletes in ancient Greece were adjured to include fennel in their diet to control their weight and the ancient Greek name for fennel, *marathon*, meaning to 'grow thin' reflects this use of the plant. In *The Good Housewife's Jewell* (1585), T. Dawson gives us a quaint recipe:

FOR TO MAKE ONE SLENDER

*Take fennel and seethe it in water, a very good quantity, and wring out the juice thereof when it is sod, and drink it first and last, and it shall swage either him or her.*

***Species:*** The main species is *Foeniculum vulgare* (golden flowers; 6ft/1.8m). There is also a dwarf variety, the Italian or Florence Fennel, *Foeniculum vulgare* var. *dulce*, which grows about 1ft/30cm high. Both are excellent culinary herbs.

***Cultivation:*** Fennel is a very hardy perennial and will grow in most locations. It requires plenty of sun in order to ripen the seeds, which may then be harvested in autumn and stored in an airtight container. Fennel is easily propagated by seed, only remember to sow them at the back of the herb patch or border because the plants will grow thick and tall. Cut leaves back after flowering each year.

### FENNEL AND HONEY SKIN REPAIR

This probably sounds odd to anyone unfamiliar with fennel's time-honored healing properties; however, this cream is invaluable for soothing sunburn, chafing or rashes.

*3 tbspns honey*
*3 tbspns white beeswax*
*2 tbspns almond oil*
*1 tbspn comfrey root powder*
*1 tbspn fennel infusion (see method)*

*Steep chopped fennel roots in boiling water and cover for 30 minutes to make an infusion; strain before use. Melt honey, beeswax and almond oil together in a double saucepan, stirring continuously. Add comfrey and fennel infusion. Pour into a lidded glass jar and shake well before each use.*

## *Freesia*

***Iridaceae*** (Iris family).

***Scent:*** Sweet, violet-like.

***Description:*** A group of bulbs with very dainty flowers and a pretty scent, freesias were first introduced to England and Europe from South Africa in the mid-nineteenth century. They rapidly became a popular favorite among florists and enthusiasts with greenhouses. More recently, however, freesia corms that are far more hardy have been developed in Amsterdam and newer varieties are available throughout the world. As a rule, the

*Xanthosoma lindeniana*

more heavily pigmented the flower, the less fragrant it will be. The yellow, cream, white and pink varieties are preferable. The purple and crimson ones are nearly scentless. Freesias are a charming sight planted in a china bowl or placed in a vine basket and then set on a sunny window sill where the warmth helps fill a room with their fragrance. They are also a pretty cut flower and will last reasonably well indoors.

***Species:*** *Freesia refracta* (cream flowers; 2ft/60cm) has the best fragrance, in my opinion. It flowers in spring and nothing is prettier than a cloud of these flowers beneath a cherry or apple blossom tree blooming at full tilt. *F. refracta* var. *alba* (snow white flowers; 2ft/60cm) and *F. refracta* (bright yellow flowers; 2ft/60cm) are also strongly perfumed.

***Cultivation:*** Freesias prefer cool, although not frost prone, conditions. They may be grown in a greenhouse to flower through winter or straight into the garden in late spring to flower the following year. Some corms are treated to flower twice and they will also bloom through to late summer. The roots should have some protection from temperature extremes. Interplanting with other fragrant, flowering bulb plants, such as *Amaryllis belladonna* can make a striking and aromatic flower bed. Another idea is to underplant small shrubby evergreen plants, such as Eriostemon or Erica, with freesias. So protected, the corms should come up after even a harsh cold snap and bloom again.

### TRADITIONAL FLORAL WASHBALLS

The Tudor court bathed in floral toilet waters, washing themselves with balls made from powdered soap and grated herbs or barks. These acted more like pumice stones than the smooth soap we know today—nonetheless, they make a pretty gift.

❖

*2½oz/75g Castile soap, grated*
*4½fl oz/125ml rosewater*
*essential oil—rose or carnation*
*1 tbspn marjoram, dried and crushed*
*1 tbspn lavender flowers, dried and crushed*
*1 tbspn rose petals, dried and crushed*
*vegetable coloring, if desired*

*Melt soap and combine with rosewater over a low flame, stirring thoroughly. Cool slightly, then add oil and crushed marjoram, lavender and rose petals; mix well. Roll into balls and leave in the sun on a piece of waxed paper for about 2 hours. Wet your hands with a little extra rosewater and polish each ball until smooth. Leave to dry completely overnight.*

## *Galanthus*

SNOWDROP

***Amaryllidaceae*** (Amaryllis family).

***Scent:*** Sweet and cool; mossy or almond-like, depending on species.

*Galanthus nivalis*

***Description:*** The dainty snowdrop received its common name because of its early spring-flowering habit, often pushing its way through late snow. Its formal name comes from the Greek *gala* and *anthos*, meaning 'milk-flower'. In Italy it is still called the Milk Flower, though the French call it *perce-neige*, a reference to its snow-hardiness. Folklore takes a dim view of the snowdrop, primarily because the closely-capped flowers are thought to resemble a corpse in a shroud. Country folk believe snowdrops are harbingers of death or ill luck, at least if they are brought into the house. Despite this, snowdrops are a much-loved garden favorite, their delicate nodding heads and pronounced perfume spreading smiles among all who see them, a reminder that 'spring is here'.

***Species:*** *Galanthus nivalis* (white flowers with

green-tipped petals; 6in/15cm) is the Common Snowdrop, native to Europe. Other pretty varieties include the almond-scented *G. allenii* (7–9in/ 18–25cm); *G. flavescens* (pale yellow-tipped flowers; 6in/15cm); *G. elwesii* and the autumn-flowering *G. regina-olgae*. There are also several double-flowering varieties, which are even more heavily scented than the single forms.

***Cultivation:*** Galanthus are hardy bulbs that spread easily in a garden. They prefer a loose, friable, moisture-retentive soil and will last best in a position of semi shade, rather than full sun. Plant great lavish sweeps of them under deciduous trees for a massed 'woodland' effect.

## *Galium*

SWEET WOODRUFF

*The woodruff is a bonny flower, her leaves are set*
*like spurs*
*About her stem, and honeysweet is every flower*
*of hers.*
*Yet sweetest dried and laid aside unkist with*
*linen white,*
*Or hung in bunches from the roof*
*of winterly delights.*

Traditional

***Rubiaceae*** (Madder family).

***Scent:*** Sweet, like new-mown hay.

***Description:*** A genus of about 80 species, native to northern Europe and found throughout much of the world. Sweet woodruff is a slow-spreading, tiny perennial, only growing to 6in/15cm high. The plants bear small star- or bugle-shaped flowers and leaves set in whorls on the stems—the ruff-like appearance of the leaves gave the plant its common name.

The scent of fresh woodruff intensifies as it dries and sachets of the leaves remain fragrant for many years. The dried leaves of woodruff have long been used in potpourri, as a strewing herb and to freshen the air. It is also a popular culinary herb, particularly for flavoring wine and tea and sweet dishes. It was once used in great quantities for stuffing mattresses, as a moth deterrent and for perfuming linen. Thomas Tusser recommended a 'swete water' made from woodruff to soothe and cleanse the complexion. Woodruff leaves were used as bookmarks in Tudor times so that dampness and bugs would not damage the valuable paper.

***Species:*** *Galium odoratum* (white star-like flowers) is a pretty choice for a ground cover, either in shade or beneath old-fashioned roses. Other varieties include *G. suberosa*.

***Cultivation:*** Remember that woodruff is originally a woodland plant and you will rarely err in growing it. Set plants in partial shade or around trees or shrubs, ensuring they are protected with a generous dollop of leaf mold. They also prefer to be well-watered. Being tiny, woodruff is appropriate for planting alongside a path. The plants may be grown in seed trays in spring, thinned and then placed in the garden. Propagation is via root division and, once established, woodruff will spread happily by self-sowing.

## *Galtonia*

SUMMER HYACINTH OR CAPE HYACINTH

***Liliaceae*** (Lily family).

***Scent:*** Very light and sweet.

***Description:*** These bulbs are native to South Africa, specifically the Cape of Good Hope, so explaining one of their common names. Their official name was a tribute to the esteemed South African botanist, Francis Galton. Galtonias are also known fittingly as Spire Lilies, and they have been known to grow up to 5ft/1.5m tall. Closely related to the hyacinth, galtonias resemble them in appearance, bearing drooping spikes of pure white or greenish white bell-shaped flowers each summer. They make a marvelous display when massed in a border with other bulbous plants, such as liliums and iris.

▶ A touch of classical elegance can enhance the sensuous beauty of a scented garden.

***Species:*** There are only two species in this group: *Galtonia candicans* (white flowers; 4-5ft/1.2-1.5m) and the smaller *G. princeps* (greenish white flowers; 2ft/60cm).

***Cultivation:*** Galtonias are hardy plants and may be planted, either in clumps in the garden and allowed to naturalize, or into formal borders. They require well drained soil and plenty of sun in order to flower, but they can stand some neglect provided they are included in the annual routine of thorough mulching and top dressing. Try a cluster of galtonias by the letter-box or the side of a shady path where they will be brushed against.

### SCENTED BATH BAGS

Many scented herbs and flowers may be used in the bath for their aromatic and relaxing effects. Rather than putting them directly into the bathtub—where they will stick to your skin and clog up the plug hole—it is better to spoon herbs or flowers into a bath bag and add it to the bath as the water is running. Be sure to spoil yourself as you soak in the bath—put a pillow behind your shoulders and have a large soft towel with which to dry yourself.

*3 tbspns lemon balm*
*1 tbspn rosemary*
*1 tbspn chamomile flowers*
*2 tbspns bran*
*towelling face cloth*

*Place all the ingredients in the center of the cloth. Gather up the sides to make a pouch and secure firmly with ribbon. Before getting in, wet the bag and squeeze it to release the milky bran and scented essences. Use the bag to scrub yourself gently all over, thus perfuming the skin.*

❖ ❖ ❖

## Gardenia

*For the most part, fragrant flowers are light in colour and white . . . Flowers of thick texture are often heavily scented—the Magnolias for instance, Gardenias and those of the Citrus tribe.*

Louise Beebe Wilder, *The Fragrant Path*, 1932

***Rubiaceae*** (Madder family).

***Scent:*** Heavy, rich and very sweet.

***Description:*** A large group of evergreen trees and shrubs bearing velvety milk-white flowers, which fill a garden with their rich fragrance from mid to late summer. Gardenias are famed for their luxuriant flowers and they are the sentimental favorite for bridal bouquets, buttonholes, debutantes' nosegays and corsages.

Gardenias are lovely in the garden or in a greenhouse or conservatory setting. One of the most exquisite indoor arrangements I have ever seen was a potted standard double *Gardenia augusta* that had been clipped into a topiary 'ball' shape, with a mat of creeping *Gardenia a.* 'Radicans' nestling at its base. Both were in full flower—a truly lovely sight and gift to the senses.

***Species:*** Gardenia species are found throughout the world. *Gardenia augusta* (white flowers; jasmine scented; 3-4ft/1-1.2m) is native to Florida and China where the petals are used to add flavor and scent to tea. There are also the gorgeous *G. lucida* (white, pale yellow flowers, lily scented; 15ft/4.5m); *G. coronaria* (white flowers; balsamic scent; 20ft/6m) and *G. arborea* (white flowers; 20-30ft/6-10m). Smaller shrubby varieties include the dwarf *G. citriodora* and *G. augusta* 'Radicans', which are both ideal as pot specimens.

***Cultivation:*** Gardenias are not over tender, but they can be temperamental. Their soil should be lime free and they should be set in a sunny aspect. High humidity is required for flower buds to open; if you have been plagued by unopened buds dropping for no apparent reason, try misting the plant regularly. If the gardenia is potted it needs regular watering but do not let the pots sit in water. Gardenias are easily propagated by taking cuttings. Pinch out tips of young shoots to encourage bushiness.

*Gardenia* sp.

## Genista

MOUNT AETNA BROOM, SPANISH GORSE

***Fabaceae*** (Pea family).

***Scent:*** Sweet and refreshing; like vanilla or pineapple, depending on species.

***Description:*** *Genista aetnensis* is known as Mount Aetna Broom and *G. hispanica* as Spanish Gorse, because they grow so profusely in these locations. With their bright green stems, tiny leaves and clusters of 'winged' yellow flowers, genistas look particularly lovely planted next to a blooming lilac or tumbling over low walls and fences. Their sweet and pervasive scent spreads throughout the garden, adding to the wealth of summer fragrances. Genistas are a pretty focal point in a shrub border; or, they can be planted against a brick wall to mask it.

***Species:*** The prime species are *Genista aetnensis* (golden, pea-shaped flowers, slightly vanilla scent; 9–10ft/2.7–3m) and the squat *G. hispanica* (yellow flowers, pineapple scent; 2ft/60cm). A tall variety is *G. cinerea* (yellow flowers, sweetly scented), which grows to 8–10ft/2.4–3m in height. An unusual choice for a rock garden would be *G. monosperma pendula* or

*Genista hispanica*

Bridal Veil Broom, a weeping variety with tiny white flowers, which sprawls delightfully in a warm, dry spot.

***Cultivation:*** Genistas are deciduous shrubs and fairly hardy. They prefer a friable, well-drained soil and plenty of sun. After flowering, prune lightly to retain the plant's shape, but do not cut back severely.

## Geranium

CRANESBILL (*SEE ALSO* PELARGONIUMS)

*On the table . . . stands a little saucer with precious, sweet-smelling Geranium leaves . . .*

Mrs Earle, *Potpourri from a Surrey Garden*, 1905

***Geraniaceae*** (Geranium family).

***Scent:*** Lemony, foxy, distinctive.

***Description:*** Cranesbill, the 'geranium' of the Middle Ages, was first listed in Mayster Jon Gardener's *The Feate of Gardenynge*, the earliest known original treatise on growing herbs, plants and flowers. The formal name was derived from the Greek *geranos*, meaning 'a crane', a reference to the plant's 'branched' carpels. With its dainty leaves and flowers, cranesbill is a popular plant for edging and borders in old-fashioned 'knotte' gardens. In addition, the leaves have a high tannin content and were used as an astringent tonic and gargle. The Wild Cranesbill, or Herb Robert, is thought to be named for the eleventh century monk who founded the Cistercian order in England.

***Species:*** *Geranium maculatum* (mauve/purple flowers; 4-6in/10-15cm); *G. robertianum* (pink flowers on red stems; 4-6in/10-15cm).

***Cultivation:*** Cranesbill is a perennial plant and will grow well in most conditions. In many parts of England it is a wild hedgerow plant. It flowers over several months during summer and should be cut back in the garden fairly regularly as it tends to straggle.

## *Gladiolus*

***Iridaceae*** (Iris family).

***Scent:*** Clove-like, or violet-like.

***Description:*** Forget the luridly-colored and scentless varieties tortured into formal flower arrangements, and look for the old-fashioned gladiolus instead. Gladioli were native to South Africa and were first introduced to Europe with freesias by John Bartram about 1730 as a result of the Dutch colonization of South Africa. They were known as Corn Flags or Sword Lilies and, although desirable for their perfume, gladioli took so enthusiastically to English gardens that John Parkinson growled '. . . it will choke and pester [a garden] . . .'

*Gladiolus tristis* was cultivated in the Chelsea Physic Gardens by 1745 and much admired for its pale lemon-yellow flowers, which had a heavy clove-like scent. Compared with the modern derivations, these gladioli are very small, but just one sprig of the flowers will scent a whole room.

***Species:*** Very few existing species of *Gladiolus* possess perfume, but look for the following: *Gladiolus tristis* (yellow flowers with dappled red spots, releasing rich clove scent in the evening; 18in/15cm) also known as the Yellow Marsh Afrikaner or *G. alatus* (red flowers; 12in/30cm).

***Cultivation:*** Gladiolus are fairly tender bulbs and are really a hothouse plant in all but the most temperate areas. The corms should be planted in the autumn and be well protected through the winter with a thick covering of leaf mulch. When cultivated for commercial purposes, gladioli are usually either grown under glass or capped with special protective cloches to ensure the blooms stay unbruised.

## *Gordonia*

***Theaceae*** (Tea family).

***Scent:*** Sweet, light, tea-like.

***Description:*** A group of evergreen shrubs and trees native to China and Formosa. Closely related to camellias, gordonias have similarly leathery, smooth and shiny green leaves and bear a profusion of cupped, camellia-like flowers with a thick cluster of orange stems in the center of each. When all the flowers are out, they look like many fried eggs, sunny-side-up. A gordonia is a striking plant and deserves center stage in a garden for its visual and aromatic impact.

***Species:*** *Gordonia axillaris* has large white flowers (3-5in/7.5-13cm across) with golden anthers in the center and a light tea-like perfume. It flowers from late winter through to early spring. *C. chrysandra* has slightly smaller creamy flowers (3in/7.5cm across) and also blooms through winter into summer. *G. lasianthus* is a summer-flowering American variety with 4in/10cm wide flowers.

***Cultivation:*** Gordonias require an acid, humus-rich soil and will not tolerate severely cold or wet weather. Full sun, plenty of deep watering of the roots during summer and thorough mulching will result in a thriving plant. Gordonias are easily propagated via cuttings taken in mid to late summer.

*Gordonia axillaris*

APHRODISIAC POTPOURRI

A warm and dusky 'mood-setter'.

❖

*9oz/250g red rose petals*
*4½oz/125g jasmine flowers*
*4½oz/125g orange flowers*
*1 tbspn sandalwood powder*
*1 tspn cinnamon powder*
*2 tbspns dried orange zest*
*2 tbspns orris root powder*
*essential oils—musk and ylang ylang*

*Mix all the ingredients together well in a large china bowl; add extra oil as desired.*

## *Grevillea*

***Proteaceae*** (Protea family).

***Scent:*** Honeyed.

***Description:*** An interesting group of shrubs, native to hot areas that experience a long dry season each year, such as Australia, New Zealand and some parts of Africa. The red, white and rosy pink flowers are brimful of nectar and will attract bees and, in turn, honey-eating birds to a garden. The foliage is green and much divided, almost fern-like, while the flowers are dainty with curled outer petals and crimson to red bracts. Grevilleas are an ideal hedging plant.

*Grevillea* cultivar 'Sandra Gordon'.

***Species:*** *Grevillea rosmarinifolia* (red flowers; narrow emerald leaves, rather like those of rosemary; 6–7ft/1.8–2m) and *G. leucopteris* (white flowers that exude a strong perfume at night) are both good choices for a hedge or windbreak or for planting against a sunny wall. Grevilleas with a decidedly spreading habit are the white-flowered *G. triloba* and *G. australis.* They look most attractive in a rockery.

***Cultivation:*** Grevilleas are quite hardy, preferring a sandy, well-drained soil and full sun. An established plant will bear many dozens of flowers in a season.

## *Hamamelis*

WITCH HAZEL, WINTERBLOOM

*And the Witch Hazel, Hamamelis Mollis;*
*That comes before its leaf on naked bough,*
*Torn ribbons frayed, of yellow and morron,*
*And sharp of scent in English frosty air.*

Vita Sackville-West

*Hamamelis mollis*

***Hamamelidaceae*** (Witch Hazel family).

***Scent:*** Sweet, musty or yeast-like.

***Description:*** A group of fragrant trees or shrubs that flower from autumn to winter, witch hazels provide color and delightful scent during dark, cold days. The bark and leaves are much used in herbal medicine for their astringent and styptic properties. Witch hazel extract, primarily derived from *Hamamelis virginiana,* was Grandma's general household remedy for burns, scalds, cuts and all inflammatory skin conditions. It is still in general use as a cosmetic. In the traditional language of flowers, witch hazel means 'a spell', referring to its use by magicians and wise men as a means of water divination.

***Species:*** The witch hazels are a pretty and unusual choice in a scented garden. The different species have the odd habit of dropping their golden leaves in late autumn before the flower buds open. Two of the best known are *Hamamelis virginiana* (small tree, spidery yellow clusters of flowers) and *H. mollis,* the Chinese Witch Hazel, which is slightly smaller with yellow flowers. *H. vernalis* (orange, curled flowers) makes an attractive shrub to set by a stone bench or perhaps a garden seat that catches a little winter sun. *H. japonica* is an ideal potted plant with its neat, compact habit and pretty lemon-yellow blooms. For a northern Christmas, try setting a pair to either side of the front doorway as a warming and fragrant Christmas welcome. With all species, the ribbon-like flowers are followed by black nuts containing white seeds. These ripen the following summer when they are forcibly ejected from their capsules with quite a loud noise—explaining the origin of the plant's country name, Snapping Hazelnut.

***Cultivation:*** In its natural state, witch hazel grows wild in damp areas within the North American and Canadian woods. It is able to tolerate icy weather but requires sunshine and an open aspect in order to flower. So that a witch hazel can make the maximum impact in a winter garden, set it against a backdrop of evergreen trees, like yews or cypress, so the flowers will be set off to their best advantage. Witch hazel sprigs also make a lovely and long-lasting cut flower for indoors during winter.

*To make your own 'swete water' to bathe the complexion, shake together 1 tbspn of chopped rosemary with equal parts (4fl oz/125ml each) of rosewater and distilled witch hazel extract. Strain, bottle and chill before applying to face and neck with a moistened cotton wool ball. This recipe has a marked tonifying effect and will help reduce the excessive ruddiness caused by 'thread' veins. It also makes a soothing and cooling eye bath.*

## Hedychium

GINGER OR BUTTERFLY LILY, INDIAN GARLAND FLOWER

***Zingiberaceae*** (Ginger family).

***Scent:*** Sweet, heavy and pronounced.

***Description:*** Popularly known as Ginger or Butterfly Lilies, this group of tuberous-rooted bulbs are among the most beautiful and exquisitely scented plants in the world. They grow quite quickly and within the first year will sport either pure white or lemon yellow flowers with bright red filaments—an exciting and dramatic sight for a summer balcony or patio. Ginger lilies are intensely fragrant and their perfume is strongest in the evening when it can pervade an entire garden. Root fibers of one species, *H. spicatum*, were once used by the Hindus to produce incense for religious ceremonies and this product was an important commercial item for many centuries. The flowers are used by Indian women to decorate their hair and an essential oil derived from the roots is used to scent clothes.

***Species:*** For the most marvelous fragrance, try to obtain the alluring Hawaiian Lei Flower, the pure white *Hedychium coronarium. H. gardnerianum* (jasmine-scented yellow flower spikes) is also stunning. They both reach 4-6ft/1.2-1.8m and have an alluring scent even when wilting.

***Cultivation:*** Hedychium requires rich soil and plenty of water and light. They need to be over-wintered in a frost-free greenhouse during a northern winter until they come into bloom in early summer. Good drainage is not essential but free-circulating air is; these plants dislike being closed in by others and are best as free-standing specimens. Hedechium is easily propagated by splitting up and striking rootstock in spring.

## Helleborus

HELLEBORE, CHRISTMAS ROSE, CHRISTE HERBE

*Borage and Hellebor fill two scenes,*
*Sovereign plants to purge the veins*
*of melancholy and cheer the heart*
*of these black fumes which make it smart . . .*

Robert Burton, *Anatomy of Melancholy*, 1621

***Ranunculaceae*** (Buttercup family).

***Scent:*** Sweet and powerful. (NOTE: poisonous varieties smell most unpleasant.)

***Description:*** The generic name for these plants, *helleborus*, is derived from the Greek *elein* ('to injure or harm') and *bora* ('food'); in short, if you are planning to include any of these plants in a garden, be aware that many varieties are extremely poisonous! Warriors of ancient Gaul would rub the tips of their arrows with juice from the flowers, so rendering their chances in battle more favorable. It is also important to note that the most beautiful hellebore, *Helleborus foetidus* with its green and purple cup-shaped flowers, is not only poisonous but emits a most appalling smell reminiscent of decaying meat—presumably in order to warn plant-eating animals that it is toxic. Black Hellebore or Christmas Rose, *H. niger*, and the Sweet Hellebore, *H. odorus*, are better choices for a sweetly scented garden. The Christmas Rose has been a particular favorite of gardeners since Roman times; Gerard wrote:

▶ With its drooping racemes of pea-like flowers, wisteria lends a delicate air to a garden setting.

*It floureth about Christmas, if the winter be mild and warm . . . called Christ herbe this plant hath thick and fat leaves of a deep green colour, the upper part thereof is somewhat bluntly nicked or toothed . . . It beareth Rose-coloured flower on slender stems, growing immediately out of the ground, an handbreadth high, sometimes very white and oftimes mixed with a little shew of purple . . .'*

A potent cathartic, the Christmas Rose was used as a purgative and a cure for insanity in earlier times. Parkinson, in 1641, alludes to this medicinal application (which, of course, is not recommended for use today):

*. . . a piece of the root being drawne through a hole made in the eare of a beast troubled with cough or having taken any poisonous thing cureth it if it be taken out the next day at the same houre . . .*

In addition to being a certain cure for madness, the ancients believed that strewing their apartments with dried hellebore flowers would drive away evil spirits.

***Species:*** As discussed, avoid *Helleborus foetidus,* the aptly named Stinking Hellebore, and look instead for the Sweet-Scented Hellebore, *H. odorus* (greenish white drooping flowers; 18in/45cm) or the Christmas Rose, *H. niger* (white or pink flowers; 18in/45cm).

***Cultivation:*** Hellebores prefer cool, partly shaded conditions and will thrive in most kinds of ordinary garden soil. They may be readily propagated by root division the first or second summer after they have flowered but be sure to wear gloves when undertaking this task—the juice from the roots is a violent irritant to the skin.

Elegant lupins have long been a favorite in cottage gardens.

## *Hesperis*

DAMASK VIOLET, SWEET ROCKET

***Brassicaceae*** (Mustard family).

***Scent:*** Clove-like.

***Description:*** These evocatively-named plants are night-scented perennials to enjoy on warmly sweet summer nights. They are small and very pretty plants with white, purple or pale mauve flowers on long drooping spikes, which appear in early summer. Their scent is similar to that of wallflowers or stock. Gerard wrote that *Hesperis matronalis* had '. . . great large leaves of a dark green colour, snipt about the edges; . . . the flowers come forth at the top of the branches, like those of the Stock Gilloflowre, of a very sweete smell'.

Hesperis varieties, commonly known as Sweet Rocket, originated in southern Europe and Russia and have long been a favorite in cottage-style English gardens either naturalized in a woodland setting or as a massed display in herbaceous borders. Once you have it in a garden, Sweet Rocket continues to self-sow indefinitely, spilling over into neighboring flower beds and even into the lawn if unchecked. It is an appropriate choice for a 'wild' garden look, and may also be potted and brought indoors to scent rooms.

***Species:*** Also known as the Dame's Violet or Damask Violet, the most common Sweet Rocket is *Hesperis matronalis* (white/violet flower spikes; 2-3ft/60-90cm). *H. tristis* (cream or purple flowers; 18in/45cm) is a woodland plant in Southern Europe through to the Caucasus on the Black Sea. They are not as common a nursery line as *H. matronalis,* but are worth seeking out for their delicious evening-intense fragrance. Some double forms are also available.

***Cultivation:*** Hesperis varieties are very hardy and will grow in most soil types, preferring a light, sandy mixture in full sun. Cuttings may be readily taken but are really not necessary as the plants do self-seed most enthusiastically.

### PERFUMED PADDED DRAWER LINERS

These are made by cutting fine organdy or taffeta rectangular pouches to fit the base of your drawers. Stitch on 3 sides, and fill sparingly with this fragrant mixture:

❖

*2oz/50g lemon verbena*
*1oz/25g peppermint*
*25g lavender flowers*
*dried zest of 1 lemon, crushed*
*1 tspn ground nutmeg*
*1 tbspn orris root powder*

*Stitch the fourth seam. Placed in each drawer, these padded liners will protect and perfume your clothes as well as the room.*

## *Hoya*

WAX FLOWER

***Asclepiadaceae*** (Stephanotis family).

***Scent:*** Sweet, honeyed, delicate but penetrating when in full flower.

***Description:*** A group of evergreen climbers from Australia, China and the Pacific Island with a vigorous twining habit, it is not uncommon for hoyas to produce strands from 10-20ft/3-6m in length! They are, therefore, ideal for growing in a conservatory or greenhouse and may be easily trained along a permanent framework of bamboo or wire. The star-shaped flowers are extremely dainty and remarkable in their almost cosmetic perfection. Sir John Hooker wrote that the *Hoya bella*'s flower was '. . . like an amethyst set in frosted silver'. This aptly describes its appearance, which resembles pearly wax stars with a glittering 'jewel' of nectar in the center of the rose-red stigmas. Before taking the secateurs to a hoya, though, remember that if you cut the flowers you

The climbing rose, 'Titian'.

sacrifice the growth spur that provides more flowers, for there is no proper stem.

***Species:*** *Hoya carnosa* (pink, five-pointed starry flowers; 12ft/3.6m) is the sentimental favorite. Also keep an eye out for the creamy white *H. australis,* blush-pink *H. bella* with its striking mauve stigmas; and some of the newer varieties. Most hoyas bloom from late spring to late summer but some of the new varieties bloom at different times; with judicious selection it is possible to have one or another in flower through the year.

***Cultivation:*** Hoyas are happiest in a pot; in fact, they prefer to be slightly pot-bound. A rich, fibrous potting mix should be used and plants set in partial shade. They need good drainage so work some sand or grit through the potting mix first of all and allow them to dry out between waterings. Water more in summer than in winter. A little dilute liquid fertilizer during spring will help boost flowering considerably.

## *Humea*

***Asteraceae*** (Sunflower family).

***Scent:*** Incense-like, rich.

***Description:*** Gardeners have long been fascinated by slightly exotic, even temperamental, plants. Humea, a biennial that is native to parts of Australia, is one of these. Even in its home country it is happiest in a glasshouse or indoors, set against glass. All parts of the plant are scented, described by Roy Genders as having '. . . the rich smell of dry Virginia tobacco leaf . . . specially pronounced after a shower during warm weather.' Humea flowers in late summer and is well worth cultivating, for the scent is intoxicating. It is also an attractive shrub in its own right, having a neat, compact appearance and reddish feathery flower heads.

***Species:*** *Humea elegans* (reddish flowers; 4–8ft/1.2–2.4m) befits its name. A very popular plant during the Victorian era, potted humeas were set by doors adjoining ladies' boudoirs to the conservatory, so the scent would waft in to perfume their rooms at any time of the day they cared to take a stroll. Gardeners were instructed to spray the leaves and flowers regularly as they become more fragrant when moist.

***Cultivation:*** Humea does have the reputation of being difficult to grow, so buying an established plant from a reputable nursery may save a lot of frustration. However, success is to be had by sowing seed in summer, potting the new seedlings—adding a top dressing of sand—and watering very sparingly until transferring to larger pots the following summer. A layer of well-rotted manure in the final potting mix will not go astray.

*Humulus lupulus* 'Aureleus'

## *Humulus*

HOPS

*Hops transmuted our wholesome ale into beer, which doubtless much alters its constitution. This one ingredient, by some suspected not unworthily, preserves the drink indeed, but repays the pleasure in tormenting diseases and a shorter life . . .*

John Evelyn, *Pomona*, 1670

***Cannabidaceae*** (Hemp family).

***Scent:*** Aromatic, yeasty.

***Description:*** Despite the dour warnings of John Evelyn above, the pretty heart-shaped leaves and drooping flowers ensure this climbing, hardy, herbaceous plant a place in any scented garden. Thomas Tusser, in his *Five Hundred Points of Good Husbandrie* (1573) included 'hop, set in Februarie' along with 'Red myntes, cowsleps and paggles, Winter Savery, Bawlme . . . Bassell . . . and Daisies of all sortes' as excellent strewing herbs. The name 'hop' is derived from the Saxon *hoppan*, meaning to climb, and it does indeed make a wonderful summer cover for any pergola or trellis. It was also a popular choice for growing over an arbor in Shakespeare's day, because of its associations with romantic love. Many thought that the smell of ripe hops acted as an aphrodisiac. Other uses saw the seed heads being included in the brewing of beer and the dried flowers being used to stuff pillows and mattresses to induce sound sleep. George III would not go to sleep without his hop pillow. Dried hops have a sweeter smell than the fresh flowers and do have a very slightly narcotic effect. Elizabeth Yandell wrote at the turn of this century, describing her summer holiday:

*We made no end of sleepy-pillows; every one of the beds had one, a large linen bag filled with a proper quantity of hops and dried cowslips, dried in their season.*

Cowslips are soft and sweet, but a slightly more aromatic mixture for her 'sleepy pillow' would include 2 parts hops to 1 part each of dried woodruff and southernwood plus a few crushed bay leaves for tang.

Hop 'bitters', an alcoholic drink made by brewing hops with white bryony leaves and other herbs, was once drunk by farm workers at their midday meal, along with the traditional English 'ploughman's lunch' of bread, cheese and pickled onions or cucumbers. Hop tea may be taken to calm the nerves, but the taste is not to everyone's liking.

***Species:*** The Common Hop, *Humulus lupulus* (golden leaves, green/yellow flowers; to 20ft/6m and more, depending on age), is a striking and hardy climber. Also attractive are the Golden Hop, *H. lupulus* 'Aureleus', and the Chinese Hop, *H. japonicus*.

***Cultivation:*** Being a hardy plant, hops will flourish in most garden conditions, preferring a position in full sun. They are especially attractive when trained with another climber, such as clematis, providing a spectacular display against a blue summer's sky.

### SWEET HOP SACHETS

And an old piece of gardening folklore . . . a hop vine that grows in a very twisted or contorted shape should be cut, dried and hung over the front door as this is very lucky indeed.

*1 cup hops*
*½ cup mint*
*½ cup sweet woodruff*
*½ cup agrimony*
*1 tspn southernwood*
*small muslin or cotton bags*
*ribbon for tying*

*Crush the herbs slightly and mix together in a china bowl. Spoon the mixture into bags and tie securely with ribbon. Stitch ribbon loops so they will not slip and hang on a bedpost or over a lampshade where the scent will permeate the room.*

## *Hyacinthus*

*If of thy mortal goods thou art bereft,*
*And from thy slender store two loaves alone to thee are left,*
*Sell one, and with the dole*
*Buy hyacinths to feed thy soul.*

Sadi, *Gulistan: Garden of Roses*

***Liliaceae*** (Lily family).

***Scent:*** Sweet, balsamic, penetrating.

***Description:*** Percy Bysshe Shelley dearly loved the hyacinth:

*. . . the hyacinth purple, and white, and blue,*
*which flung from its bells a sweet peel anew*
*of music so delicate, soft and intense,*
*It was felt like an odour within the sense.*

These softly perfumed spring-flowering bulbs are native to southeastern Mediterranean regions. The flower derived its name from the youth Hyacinthus who was beloved of the sun god, Apollo. Hyacinthus was accidentally killed during a game of quoits and the broken-hearted Apollo decreed that he should become a flower. As he spoke, the spilled blood on the ground became a patch of exquisitely scented hyacinths.

Hyacinths have a heady, sweet perfume. They may be placed in outdoor beds or allowed to

'The hyacinth purple, and white and blue.'

naturalize under trees or in drifts in the far reaches of the garden. They can also be brought indoors in special bulb jars or pots where they present a dramatic and showy picture. There is a wide color range, including white, blue, purple, pink, plum and yellow. A charming piece of folklore advises that growing pink hyacinths will bestow amiability and goodwill upon household members.

***Species:*** *Hyacinthus orientalis,* or the Common Hyacinth (6–9in/15–25cm; white, yellow, mauve, pink, blue or purple flowers) is an ideal choice for window boxes, borders, paths or for strewing through a gravel driveway bed. This variety has given rise to the many hybridized Dutch 'florist's' hyacinths, including some lovely large double-flowered forms. As with most fragrant plants, the white hyacinths are among the most heavily perfumed, although some of the blues and lighter pinks are unsurpassed for fragrance. Try a selection of the following in a raised bed or container: 'L'Innocence' (pure white flowers); 'La Victoire' (rose pink); 'Royal Mulberry' (dark purple); 'Ben Nevis' (double white flowers); 'King of the Blues' (dark blue with purple splashes) and 'Ostara' (clear, pure blue). For sheer drama, interplant the rich, red and fragrant 'Scarlet Perfection' with the unusual golden apricot flowers of 'Orange Charm'.

***Cultivation:*** Hyacinths prefer a cool climate and will thrive either massed outdoors beneath ornamental trees or in a pot on a table in a coolish room. They are quite hardy and lend themselves to forcing for a winter flowering. One tip: bulbs may become moldy if kept in cool storage and prevented from rooting for too long; buy fresh bulbs each year if planning indoor container growing.

## *Hypericum*

ST JOHN'S WORT

***Hypericaceae*** (St John's Wort family).

***Scent:*** Resinous, balsamic; not all species have a pleasant fragrance.

***Description:*** A group of hardy, herbaceous perennials that grow wild throughout Britain, Europe and Asia. The flowers are a cheery bright yellow and make a wonderful ground cover, especially for difficult positions such as driveways, where other plants may refuse to grow. The name *Hypericum* is derived from the Greek meaning 'over an apparition', a reference to its use in mystic rites, when it was employed to repel evil spirits. St John's Wort was also used widely for its medicinal action, being a potent nerve tonic, astringent and expectorant. A tea made from the leaves (1oz/25g of the leaves to a pint of water) is useful for catarrh and other lung disturbances. This infusion has also been found effective in countering bedwetting in children. John Gerard described an 'oil of St John's herb' made by infusing the herb in olive oil, which would help cleanse wounds 'made by a venom'd weapon'.

St John's Wort, said to be named for St John the Baptist, also has pretty love associations. Country folk tell that if a girl picks a sprig on St John's Day (June 24), places it beneath her pillow that evening and finds it still fresh in the morning, she could be reasonably sure of marrying within the year. The Common St John's Wort, *Hypericum perforatum* (which actually has an unpleasant, goat-like smell) blooms on or around St John's Day each year and the leaves are said to be the blood of the martyred saint.

***Species:*** The beautiful Rose of Sharon, *Hypericum calycinum*, may be planted as a ground cover and enjoyed for its sunny buttercup-yellow flowers, although it is almost scentless. *H. perforatum* and *H. prolificum* should be avoided in a scented garden as they have an unattractive aroma, reminiscent of wet fur and goats. A more attractive choice is *H. androsaemum* or the Woodland Sweet Amber (golden flowers; 3-4ft/1-1.2m), which has a most pleasing balsamic aroma. This variety was widely cultivated in medieval French monastery gardens where it was known as *toutsaine* or 'all-heal' because of its medicinal powers.

***Cultivation:*** St John's Wort is very hardy and will grow easily in most locations. To thicken appearance and encourage flowering, prune hard —to ground level, when plants are well established—in early spring.

## *Hyssopus*

HYSSOP

A WATER TO CAUSE AN EXCELLENT COLOUR AND COMPLEXION

*Drink six spoonfuls of the juice of Hyssop in warm Ale in a Morning and fasting.*

From *The Receipt Book of John Nott*, Cook to the Duke of Bolton, 1723

***Lamiaceae*** (Mint family).

***Scent:*** Honeyed, sweet.

***Description:*** 'Pretty and sweet' said John Parkinson of this hardy evergreen herb in 1640 and, indeed, it still offers the scented garden a very sweet fragrance and most decorative appearance. Hyssop has been an important religious and medicinal herb since biblical times, deriving its name from the Hebrew *azob*, meaning 'holy plant'. At the crucifixion of Jesus Christ, a vinegar-soaked sponge was wrapped around a branch of hyssop before being handed to the Savior's dying lips. Hyssop was also used widely in Jewish purification and sacrificial rites, for example, a piece of hyssop sprinkled with the blood of the Paschal lamb was used to mark doorposts when the Jews commenced their great exodus from Egypt.

Hyssop was much used by the Elizabethans in their formal gardens and mazes were usually 'sette with isope'. Thomas Tusser mentions it as one of the strewing herbs and it was also included in the 'tussie-mussies' carried during times of plague and pestilence to ward off infection. Spenser wrote that 'Sharp Isope' could be used to heal wounds. The young tops and flowers were used in herb 'pottages' and 'sallets'; hyssop tea, cordial and honey were all famous for their pectoral and carminative properties, as well as their undeniably delicious flavor. Hyssop is a most attractive garden shrub, with a bushy habit. It is a good companion plant for roses as well as other herbs and Thomas Hyll wrote in *The Arte of Gardenynge* (1564) that '. . . it [marjoram] may either be sette with Isope and Time, or with winter savoury and Time, for these endure all the winter thorow greene.' Hyssop is useful about the house and may be used to perfume rooms and give linen a clean, fresh scent, to scent candles and notepaper, and to add color and fragrance to potpourri. The flowers can be ground and used as an effective tooth powder or for refreshing aromatic baths, which also have a positive effect on muscular stiffness or aches. An old seventeenth century manuscript *Arcana Fairfaxiana*, contains a recipe, 'To Make a Bath for Melancholy', as follows:

*Hyssopus officinalis*

*Take Mallowes, pellitory of the Wall, of each three handfulls; Camomell flowers, Melilot flowers, of each one handful; hollyhocks, two handfulls; Isop, one great handfull, Senerick seed one ounce, and boil them in nine gallons of Water until they come to three, then put in a quart of new milke and go into it bloud warm or something warmer . . .*

***Species:*** *Hyssopus officinalis* (mauvish flowers; 2ft/60cm) or the tiny *H. officinalis* (blue flowers; 9in/25cm) are both charming evergreens that do well in garden nooks, by paths or tucked into herbaceous borders.

***Cultivation:*** Hyssop is hardy and will grow even in quite sandy soil; it requires full sun to flower and is much loved by bees. Hard pruning is essential to avoid legginess.

## *Iberis*

CANDYTUFT

***Brassicaceae*** (Mustard family).

***Scent:*** Very sweet and penetrating.

***Description:*** 'Candytuft' is the country name given to several species of small shrubby annual or biennial plants. It is derived from *candia,* the name for ancient Crete, which is where the plants were once most plentiful. Most varieties have rounded heads of white or rosy-pink flowers that bloom in early summer, forming broad carpets of leaves and flowers. Several varieties, such as the dark red *Iberis umbellata* and the evergreen *I. sempervirens* are scentless; however, they create a pretty and colorful effect when sown with their fragrant, albeit colorless or pale, cousins, such as *I. amara* and *I. odorata.*

***Species:*** *Iberis amara* or Rocket Candytuft (large—to 2in/5cm long—white flowerheads; 12-18in/30-45cm) is hardy and suitable for planting under windows or by doorways where it will flower, even in part shade. It is also a pleasing addition to annual borders or may be potted for decorating an outdoor patio or balcony.

***Cultivation:*** Candytuft is quite hardy and will thrive in part shade or sun in most soil types. Sow seeds in shallow drills in punnet boxes in early autumn, then prick out seedlings and set in desired position in the garden. Alternatively, sow direct into the garden for a summer blooming the following year. Candytuft may be treated as a biennial and sown in late winter/early spring as well as autumn, thus providing a longer flowering period from spring through summer.

*Iberis sempervirens*

## *Indigofera*

***Fabaceae*** (Pea family).

***Scent:*** Vanilla.

***Description:*** A group of deciduous and evergreen shrubs that are very attractive ornamentals in the garden. They are native to subtropical countries, particularly the East Indies, India, Korea and China, but most are hardy and will do well as potted standards in a warm, sheltered position. The different varieties have long, drooping, wisteria-like racemes of rose, purple or white flowers and feathery, acacia-like foliage. The famous blue dye, indigo, was once obtained from various species of indigofera, notably *Indigofera pseudotinctoria*. Indigo is still a very important commercial product for some countries; it is also used medicinally as an emetic.

***Species:*** Outstanding for its beauty is *I. decora* (white and crimson flowers; 1ft–18in/ 30–45cm).

***Cultivation:*** Indigofera do not tolerate frost or overly chilly winds but are, otherwise, quite undemanding. The bunches of flowers are produced continuously from early to late summer. When blooming is over, trim back dead stems and thin very lightly.

### KITCHEN SUNSHINE POTPOURRI

A tangy and energizing recipe for the kitchen.

❖

*2oz/50g hyssop*
*1oz/25g southernwood*
*1oz/25g marigold flowers*
*1oz/25g tansy buttons*
*1oz/25g yellow everlasting daisies*
*2oz/50g lemon thyme*
*1 tbspn nutmeg chips, crushed*
*1 tbspn bay leaves, crushed*
*2 tbspns orris root powder*
*essential oils—lemon or cedar*

*Combine all dry ingredients in a large china or pottery dish, add 3–6 drops preferred essential oil.*

## *Iris*

***Iridaceae*** (Iris family).

***Scent:*** Sweet; reminiscent of violets, orange-blossom or fruit, depending on species.

***Description:*** The iris is named for the Greek

▶ A dazzling array of color and scent.

goddess of the rainbow. It was also sacred to the Roman Juno and symbolized her power and beauty. The three leaves came to represent Faith, Wisdom and Valor and the flowers were held, much like a scepter, by Egyptian kings. Irises comprise a family of rhizomatous and bulbous plants that are famed for their stately and attractive flowering habit. The rhizomatous types are also popular for their scent, especially *I. pallida* and *I. germanica,* which have been used to produce orris root powder, a violet-scented fixative much used in potpourri and cosmetic preparations. Parkinson preferred, 'for his excellent beautie and raretie the great Turkie Flower de Luce . . . for a sweet powder to lay among linnen and garments to make sweet waters to wash hand gloves or other things to perfume them.' Orris root has a very ancient history; there are several references to it in the Bible and the Greeks used the powder to scent their clothes. In the time of Elizabeth I, it was used by weavers and tailors to perfume what was then described and sold as 'swete cloth'. Today, orris root powder comes mainly from Tuscany and Sienna, where it is known as *giaggolo.* The true strength of the roots does not become apparent until they are at least two years old, and their scent continues to intensify with passing time.

***Species:*** *Iris germanica,* which is also known as the Common Flag, German Iris or Blue Fleur de Luce smells rather like orange-blossom. It is one of the oldest cultivated flowers, having been listed as one of the plants present in the ninth century monastery garden of Reichenau. The flowers are deep blue with a bright yellow 'beard' and the petals are sprinkled with purple. *I. pallida* described by Gerard as '. . . exceedingly sweet, much like the orange flower' is also very fragrant and the color of the flowers varies from gray to mauve to a very pale gray, depending on the variety. *I. florentina,* which appears on the coat of arms for the city of Florence, has large white flowers with a yellow 'beard', which are very lightly splashed with lavender. Although the bulbous irises are generally less important scent-wise, it is worth asking a specialist nursery for the Persian Iris, which has a powerful violet perfume and beautiful blue-green petals splashed with darkest purple. The Russian Iris, *I. reticulata* and the beautiful golden-flowered Turkish Iris, *I. danfordiae,* are all delightful to behold and fragrant. Also look for the winter-flowering forms, such as *I. unguicularis* (syn. *I. stylosa*).

***Cultivation:*** Irises have been cultivated in gardens for many centuries and are quite hardy. They prefer a slightly limey, well-drained soil and will tolerate shade, though they need summer sunshine in order to flower. If planting rhizomatous varieties, set roots just below the surface of the soil in midsummer; after flowering has completed the following year, the roots may be lifted and separated to increase the numbers. Bulbous irises also require a sunny position to flower. Shallow-plant bulbs—3in/7.5cm deep—and divide after the first year.

TO MAKE PERFUMED POWDER

*Take four ounces of Florence Orris, four ounces of dry'd Damask Roses, half an ounce of Benjamin, a quarter of an Ounce of Storax, as much of Yellow Saunders, half a dram of Cloves, and a little Citron Peel; pound all these in a Mortar to a very fine Powder, put to them five pounds of Starch powdered. Mix them well, sift it fine, and keep it dry for use.*

From *The Receipt Book of Charles Carter,* Cook to the Duke of Argyll, 1732.

## *Itea*

***Saxifragaceae*** (Rockfoil family).

***Scent:*** Sweet, long-lasting; stronger in the evening.

***Description:*** This is a group of evergreen or deciduous trees with unusual and elegant drooping racemes of greenish or white flowers. Its name is derived from the Greek for 'willow' and the clusters of flowers are somewhat reminiscent of trailing willow catkins. Iteas are native to temperate areas, such as southern China and the eastern seaboard of the United States. They are bushy, free-flowering and quite hardy, and are suitable for outdoor cultivation in a warm sheltered position or for cool greenhouse conditions.

***Species:*** *Itea ilicifolia* (green/white drooping clusters of flowers; 6–8ft/1.8–2.4m) is a bushy evergreen which will thrive if planted by a sun-warmed brick wall. Its serrated glossy dark green leaves are almost holly-like in appearance. *I. virginica* (white flowers; 7–8ft/2–2.4m) is the deciduous variety and its leaves turn a rich, vibrant red after flowering is completed.

***Cultivation:*** Iteas need to be planted in a moisture-retentive soil as they do not like to dry out. They grow very well in swampy soil or by water, which are usually considered trouble spots in a garden. Light shade and afternoon sun are a winning combination and the flowers, which appear all through summer, are an elegant as well as fragrant asset in a garden. No regular pruning is required; iteas grow into an attractive rounded shape with very little help.

*Itea ilicifolia*

## *Ixora*

***Rubiaceae*** (Madder family).

***Scent:*** Rich and slightly exotic.

***Description:*** Ixoras are extremely striking ornamental evergreen shrubs and small trees. They are free-flowering, bearing clusters of brilliant orange-red or white four-petaled flowers. Native to tropical areas, such as Indonesia and India, ixoras require a high temperature and degree of humidity in a garden. If this is not possible, however, they are arresting hothouse specimens and will do well indoors. In India, the ixora bush is held sacred to Shiva and Vishnu; its black berries are said to be the favorite food of peacocks.

***Species:*** *I. grandiflora* (white flowers) is deliciously fragrant and flowers profusely. The striking colored varieties, *I. chinensis* (scarlet flowers) and *I. jaranica* (flame/orange flowers) are scentless, unfortunately.

***Cultivation:*** Being tropical plants, ixoras require high humidity in either a sunny garden or greenhouse cultivation. The plants are best propagated by means of cuttings; strike them in peaty compost in a shady situation until they are ready to go into the garden.

## *Jasminum*

JASMINE

*The owner also may set the jacemine tree bearing a fragrant flowre, the muske rose, damask and privet tree in beds, to shoote up and spread over the herber [arbour] . . . it yieldeth a delectable smell, much refreshing the sitters underneath it . . .*

Thomas Hyll, *The Gardeners Labyrinth* (1577)

***Oleaceae*** (Olive Tree family).

***Scent:*** Heavy, sweet, lingering.

***Description:*** The jasmines are probably one of the best known and highly esteemed groups of

*Jasminum polyanthum*

scented shrubs or climbers. In India, jasmine is revered as sacred to Vishnu and the flowers are often strung into garlands for votive offerings in religious ceremonies. In India and Borneo, women wear the flowers in their hair at all times; the flower's perfume combines with their hair oil, bestowing long-lasting fragrance.

There are many different varieties, mostly natives of the warmer regions. Queen of them all would be *Jasminum sambac,* the Arabian jasmine, and the Crown Princess, *J. officinale.* The flowers of *J. sambac* add fragrance to Chinese tea; it is also this variety that is most used for perfume extraction. Jasmine plants were introduced many centuries ago to China from India and Arabia and the flower became much loved by the Chinese. In his *Records of the Plants of Southern China,* third century historian Chi Han described Canton during the jasmine season:

> *. . . the city was like snow at night, and was fragrant everywhere. The flowers were used in making perfumes and scented oils to rub on the body. Indeed, everyone had the delicious scent about them . . .*

During the summer months jasmine exhales a heady and exotic fragrance that stirs the senses. It is little wonder that it is a popular choice for a romantic bower or garden of love.

***Species:*** *Jasminum officinale* (sprays of pure white flowers; 10–12ft/3–3.6m); *J. polyanthum* (fragrant white flowers; 10ft/3m); *J.* x. *stephanense* (pink flowers; 6ft/1.8m), and the rose-red *J. beesianum* are all beautiful climbers, which add color and fragrance to a scented garden.

***Cultivation:*** Jasmines are hardy and vigorous climbers. Most bloom throughout summer, but there are one or two winter-blooming varieties, notably *J. nudiflorum.* Jasmines produce much twiggy growth and should be trained firmly over an arbor (rather than against a wall, as it will penetrate the mortar) and pruned hard. Ordinary soil is fine for jasmine, but its roots should be well protected so plant some rosemary or santolina at the base of the plant. Jasmine looks marvellous entwined with another climber, such as variegated ivy, clematis or honeysuckle.

### JASMINE SCENTED DRAWER LINERS

❖

*Take lengths of thick parchment-style paper, floral wallpaper is ideal, and cut to fit the inside of your drawer. Cover with dried jasmine flowers, roll up lengths and seal in plastic bags. After six weeks, the paper will have absorbed the perfume. To intensify the fragrance, dab a few drops of jasmine oil onto a cotton ball and wipe round the inside of each drawer before setting paper inside.*

❖ ❖ ❖

## *Juglans*

WALNUT TREE

***Juglandaceae*** (Walnut Tree family).

***Scent:*** Resinous, refreshing, balsamic.

***Description:*** A group of trees much valued for the resin produced by their leaves, which acts as a fly repellent. In old-fashioned gardens a walnut tree was often planted by the kitchen door or to shade the chicken run for this purpose. Crushed or bruised, walnut leaves give off a very pleasant, head-clearing aroma and this may explain why the Italians recommend wearing walnut leaves in the hat to ward off headaches and/or sunstroke. Walnut oil is a popular hair tonic and it was once a favorite of artists who used it to blend colors on their palette (because it is very viscous and takes a while to set, the painter would have had time to blend a shade to exactly the color he wanted).

***Species:*** Both *Juglans nigra,* the Black Walnut, and *J. regia,* the Common Walnut, are attractive and scented—if rather large (they can reach 160ft/50m)—ornamental trees for the average garden. In medieval times, a walnut tree was often planted at the center of a walled courtyard and its branches would spread to each of the four walls, creating an aromatic and shady 'roof' in summertime.

***Cultivation:*** Native to northern climates of Britain and America, walnuts are very hardy and will tolerate some frost. They do require plenty of moisture during summer, particularly when they start to produce nuts, which is after anything up to 8–10 years.

*Juglans* sp.

*Lathyrus odoratus*

## ***Lathyrus***

SWEET PEA

*Training the trailing pea in clusters neat,*
*Perfuming evening with a luscious sweet . . .*

John Clare

***Fabaceae*** (Pea family).

***Scent:*** Sweet, vanilla-like.

***Description:*** One of the most popular annuals for spring gardens around the world is the sweet pea. They are easy to grow and a handful of the old-fashioned variety, *Lathyrus odoratus,* will fill a room with fragrance. This species, which was introduced to England from Sicily in about 1699, has purplish/maroon and blue petals. In the 1700s, Thomas Fairchild described them as '. . . flowers that will grow well in London, fit for adorning of squares, for the sweet scented pea makes a beautiful plant. The scent is something like honey, and a little tending to orange flower smell. These bloom a long time.' The flower was first named 'sweet pea' by the poet, John Keats.

Hybridists have bred many varieties for picking from the *L. odoratus*. The first Giant Frilled Sweet Pea, named in honor of the Countess Spencer, appeared at the beginning of the twentieth century and there are several other key strains: the 'Early (winter) Flowering'; the 'Multifloras'; 'Dwarf Cupids'; 'Knee Hi' and 'Bijou', the latter two being most appropriate as edgings. Colors vary from lavenders, mauves and bluey purples through to pink, salmon, orange, scarlet and then to white and cream. The bluish shades are most sweetly scented, the orange and red types least so.

Sweet Peas are primarily cultivated as cut flowers. They were popular in Edwardian times for this purpose, and were a great favorite of Queen Alexandra. They can also be used in salads as an edible decoration. In France, sweet peas are given to a bride as a lucky charm, for they will cause everybody to tell her the truth and '. . . give her steadfastness of purpose, which carries a good woman past all evil'.

***Species:*** *Lathyrus odoratus* is the best known, scented, climbing annual type. Depending on soil quality, they may grow up to 6ft/1.8m tall. Pretty varieties include 'John Ness' (mauve, clove-scented flowers); 'Crimson Excelsior', 'Geranium Pink', and the large-flowered night-blue 'Vogue'. Perennial sweet peas include the Everlasting Pea, *L. latifolius* and the hardy rose-red *L. grandiflorus*.

***Cultivation:*** Sweet peas require a rich, limey soil, plenty of sunlight and a good strong trellis to climb on. They are quite greedy and will revel in plenty of compost and fertilizer. They make a delightful scented hedge (5–6ft/1.5–1.8m high) if supplied with an appropriate climbing frame. They also look lovely trained through evergreen trees and shrubs. Sweet peas are readily grown from seed. If growing potted specimens, pinch back new growth for a bushier appearance and more flowers. And, for the superstitious, if sweet peas are planted on Good Friday, they are said to grow more rapidly.

## *Laurus*

BAY

*Bay serveth to adorn the house of God, as well as man; to procure warmth, comfort and strength to the limbs of men and women by bathings and anoyntings out and by drinks, etc., inwards; to season the vessels wherein are preserved our meates, as well as our drinkes; to crown or encircle as with a garland the heads of the living and to stick and decke forth the bodies of the dead; so that from the cradle to the grave we still have use of it . . .*

John Parkinson, *Paradisi in Sole Paradisus Terrestris*, 1629

*Laurus nobilis*

***Lauraceae*** (Bay Tree family).

***Scent:*** Pungent, aromatic, spicy; leaves are lavender-like, twigs have a wintergreen aroma, and berries are camphoraceous.

***Description:*** A group of aromatic shrubs or small trees, native to southern Europe and North America. They were much honored by the ancient

### BAY LEAF HEADACHE PILLOW

❖

*Bay leaves help clear the head and induce sleep. Fill a soft pillow with dried lavender, lime blossoms, scented geranium leaves and about a dozen bay leaves, crumbled. Give to a fractious patient suffering from headache, earache or fever.*

❖ ❖ ❖

Greeks who dedicated the trees to the god Apollo. The Romans used garlands of the leaves to crown their distinguished poets, doctors and returning soldiers. In fact, the name laurus is derived from *laudis,* meaning 'to praise'. The leaves were thought to clear the air and repel evil spirits and disease. For this reason, the Emperor Claudius tried to avoid the plague by moving his court to Laurentium, noted for its many bay trees.

Bay leaves beneath the pillow are effective against insomnia and they were used in Tudor pomanders and cushions for the same purpose. They were used for strewing in churches and homes, in restorative baths to relieve aches and pains and to flavor food and drinks. *The Goodman of Paris* (1393) describes a dessert of '. . . cooked apples and large Provencal figs roasted with bay leaves thereon' though today bay leaves are more likely to be used in preparing savory dishes and meat. They are a key ingredient of bouquets garni.

***Species:*** *Laurus nobilis* or Sweet Bay (tiny yellow flowers; 50ft/15m) is also known as the Victor's Laurel or Poet's Laurel. It is a strong evergreen and a popular choice for topiary work as it is easily clipped into a ball or pyramid shape. In the garden it may either be allowed to grow as it pleases or clipped into an attractive hedge, its flat, papery, polished-green leaves releasing a spicy aroma whenever crushed or bruised.

***Cultivation:*** Bay trees are reasonably hardy in a warm climate but should be sheltered indoors if the area is subject to winter frosts. Planted in a decorative pot of moist acid soil and placed in a warm aspect, a bay tree should flourish. Regular composting is required. Propagation is by way of cuttings taken in summer. The leaves may be harvested and dried for use in cooking or potpourri year-round.

## *Lavandula*

LAVENDER

*. . . tyed up with small bundles of lavender toppes; these they put in the middle of them to lye upon the toppes of beddes, presses, etc. . . . from the sweet scent and savour it casteth . . .*

John Parkinson, *Paradisi in Sole Paradisus Terrestris,* 1629

***Lamiaceae*** (Mint family).

***Scent:*** Sharp, clean, fresh.

***Description:*** These beautifully scented small shrubs have retained their popularity through the centuries. The gray-green foliage and fragrant flowers provide a pretty edging for paths and borders. It is also an ideal hedge plant, making it a favorite in older style cottage gardens or herb gardens and was much used in Tudor 'knotte-gardens', which featured labyrinths and mazes of low clipped hedges of lavender. The leisured nobility of the time would amuse themselves by playing hide-and-seek among the '. . . arbors and ayyes so pleasant and so dulce'; Cardinal Wolsey's garden at Hampton Court was '. . . so enknotted with Lavender, it cannot be expressed'.

Traditionally, lavender was used as a strewing herb for floors and cupboards. The dried flowers were placed in beds and oaken presses where they perfumed handkerchiefs and linen, causing Izak Walton to sigh longingly in 1653:

*Let's go into that house, for the linen looks white and smells of lavender and I long to lie in a pair of sheets that smell so . . .*

During Elizabethan times it was usual to sprinkle lavender water on the floors of houses and lavender seeds would be pounded in a mortar and the aromatic oil rubbed into the oak furniture to give a high gloss. Apart from enjoying the scent, housekeepers knew lavender was a powerful weapon against moths, fleas, silverfish and flies. Eugene Rimmel described lavender as '. . . a nice clean scent and an old and deserving favourite'; commercial perfume houses still use it as the basic ingredient of many fragrances. It may also be used to make a variety of cosmetic aids, for example, lavender is an effective mouthwash, skin tonic and eye-lotion, and helps to treat mild infections or skin irritations, especially rashes.

Lavender has long enjoyed royal patronage. Elizabeth I quaffed lavender tea to soothe her frequent headaches. She also spent extravagantly on lavender and carnation scent, including a certain compound '. . . to cleanse and keep bright the skynne and flesh and to preserve it in perfect state.' Eau de Cologne, based on citrus oils, bergamot and lavender, was Napoleon's favorite perfume, and the fashion for lavender-scented perfumes was continued by Queen Victoria. Said to obtain her lavender '. . . direct from a lady who distills it herself', a contemporary court gossip wrote that '. . . the Royal Residences are thoroughly impregnated with the refreshing perfume of this old-fashioned flower, and there is none that the Queen likes better.'

*Lavandula dentata*

***Species:*** The most common variety is *Lavandula angustifolia* (syn. *L. vera*) usually known as English Lavender, which is the source of the best quality fragrant oil. This is the best type for drying for potpourri. *L. dentata*, or Italian Lavender and *L. stoechas* or French Lavender, are also fairly common. Rarer varieties include the pink, *L. rosea*, and dwarf forms, notably *L. nana compacta*, the Munstead, which only grows 18in–2ft/45–90cm tall. There is also a white lavender, *L. angustifolia* 'Alba', with a deliciously sweet scent. John Parkinson preferred it to all the other lavenders and wrote of it that '. . . there is a kind hereof that beareth white flowers and somewhat broader leaves, but is very rare and seene but in a few places . . . because it is more tender and will not last so well.'

***Cultivation:*** If you remember that lavender originated in the hot stony fields of Provence, you are unlikely to err when growing it; in short, do not over-water or over-feed the bushes. Lavender should preferably face north or northeast and it likes an open position where the soil is fully drained. Regular hard pruning in autumn will produce flowers over a much longer period. And do not waste the prunings; strip the leaves for making potpourri and toss the branches on the open fire or barbecue to add a new flavor to cooking.

### LAVENDER NOTEPAPER

❖

*Store sheets of paper and envelopes, preferably tinted soft pink or mauve, in a box and liberally sprinkle each layer with lavender. An olde-worlde crinkled effect may be given to the paper by misting it with a spray of lavender water before drying. Decorate your notepaper by mounting a few dried lavender buds onto each sheet with transparent craft glue. If it is to be a gift, a charming accessory is scented ink—add 1 tspn of lavender essence (not oil) to a bottle of purple ink for truly deathless prose!*

## *Lawsonia*

HENNA

*Thus some bring leaves of henna to imbue*
*The finger ends of a bright roseate hue,*
*So bright that in the mirror's depth they seem*
*Like tips of coral branches in the stream.*

Thomas More

***Lythraceae*** (Loosestrife family).

***Scent:*** Heavy and rich.

***Description:*** *Lawsonia inermis* is the single species found in this genus and it is best known as the source of henna dye, which is derived from the leaves and young shoots. Native to East Africa and parts of the Middle East, henna has been used since ancient times by women to enhance their beauty. The powdered leaves are made into a paste for dyeing the hair a rich auburn, a paint made from the fruit will tint nails a reddish yellow and both an oil and an ointment to keep limbs supple can be produced from the flowers.

The Malays have a special 'henna dance', which is performed at weddings. They also use the leaves medicinally, as a warm poultice for foot complaints or, internally, as a tisane for a sore throat. The flowers have a most delicious scent and small wreaths of them were once worn about the head and neck by Hindu maidens. Henna was thought to be the 'camphire' referred to in the Bible: 'My beloved is unto me as a cluster of camphire in the vineyards of Engedi', a reference to the planting of these bushy shrubs around vineyards as protection against the searing desert winds. It was also thought to have been placed, with Gallica roses, in the ornamental, scented Hanging Gardens of Nebuchadnezzar.

***Species:*** *Lawsonia inermis* is a tropical shrub 8–10ft/2.4–3m high with panicles of small fragrant red and yellow flowers. (Local or country names may include Egyptian Privet, Smooth Lawsonia and Jamaican Mignonette, but they all refer to the same species.)

***Cultivation:*** Lawsonia is a reasonably hardy plant, requiring a very warm to hot situation and sandy, well-drained soil. It should not be subjected to excessive moisture or over-feeding, although high humidity is tolerated.

### HENNA HAIR GLOSS

This recipe leaves hair wonderfully smooth and shiny.

❖

*1oz/30g clear henna wax*
*5fl oz/150ml clear honey*
*1fl oz/25ml jojoba oil (from health food store)*

*Melt all the ingredients together and comb paste through hair. Sit out in the sun and allow to dry, then rinse out thoroughly with warm water.*

▸ Double Pink Cherry in its flowering glory.

## *Leptospermum*

TEA TREE OR TI-TREE

***Myrtaceae*** (Myrtle family).

***Scent:*** Refreshing, aromatic, lemony foliage; sweet flowers.

***Description:*** These species of shrubs and small trees are mostly native to Australia, Tasmania and New Zealand. They are highly aromatic and have been described as the 'colonial counterpart' to the English broom (*Cytisus*). Tea trees have small leaves which are leathery and tiny (¼in–1in/6mm–2.5cm across) white, pink or red flowers, which are clustered together on long stems. They are quite hardy, growing on windswept hills and dunes or exposed coastal areas. Tea trees can be grown as screening plants or hedges in the garden; they are also a lovely potted plant and will do well in a sunny, sheltered courtyard.

Their common name, tea tree, refers to the fact that early European colonists used the pungent leaves to make tea. Not only was this a most pleasant drink, it was favored by the explorer Captain Cook to help overcome scurvy among his men. The scent of the foliage is said to repel mosquitoes and people in the bush would always rub the leaves over their face, neck and hands before setting out on a trip. Plant a bush by the kitchen door or front gate, where the foliage can be readily rubbed between the fingers to release the fresh and pleasantly lemony scent. All the tea-trees are rich in nectar and produce a dark, strongly flavored honey.

***Species:*** *Leptospermum petersonii,* or Lemon Scented Tea Tree (white or pink flowers; 3ft/1m); *L. scoparium* or New Zealand Tea Tree also known as *manuka* (white, pink or crimson flowers; 3ft/1m). There are many different cultivars of *L. scoparium.* Perhaps the prettiest is 'Keatleyi', discovered by a certain Captain Keatley in 1917 when he was charting the northerly waters of New Zealand. It has quite large white flowers (up to 2in/5cm across) with plum-colored centers and is richly fragrant.

***Cultivation:*** Tea trees do best in a sunny position in sandy, well-drained soil. Regular composting with leaf-mold will help retain moisture in the soil. They do well in seaside gardens though they may assume a prostrate habit to cope with buffeting strong winds. After flowering is completed, prune the bush very lightly to avoid a leggy appearance. Be sure to keep prunings for use on an open fire during winter as the burning wood has a marvelous, cleansing scent.

### TEA TREE HAIR TONIC

A stimulating massage treatment for scalp and hair problems, especially dandruff. Diluted, this recipe will also help clear up cradle cap in babies.

*5 drops tea tree oil*
*5 drops eucalyptus*
*5 drops oil of bay*
*3½fl oz/100ml pure alcohol*
*3½fl oz/100ml caster oil*

*Mix all the ingredients together. Warm, and massage gently into scalp.*

## *Leucadendron*

SILVER TREE

***Proteaceae*** (Protea family).

***Scent:*** Soft and sweet.

***Description:*** In my opinion, this is a group of shrubs and trees that deserve to be better known. They are native to South Africa and have ivory-white flowers and silvery-toned leaves, which explains the derivation of their botanical name: from the Greek *leukos,* meaning white and *dendron,* meaning tree. The leaves overlap each other tightly, masking the dark branches, so the whole tree appears to shimmer in a light breeze. They are widely planted about Cape Town, where the leaves are pressed and sold as souvenir bookmarks and placemats.

***Species:*** *Leucadendron argenteum,* the Silver Tree (20–25ft/6–8m), has rounded golden flower heads, which emit a very light, soft perfume. *L. discolor* or Flame Gold Tips (6ft/1.8m) is an exquisite shrub and a popular spring cut flower but, unfortunately, it is not scented. All leucadendrons bear male and female flowers on separate plants. The former are the most attractive, being fluffy and bloom-like, while the female flowers are more like buds or cones.

***Cultivation:*** Leucadendrons are hardy, drought resistant and wind-tolerant. They require a compost-rich soil, preferably with a little sand worked through it. Gentle heat is required for flowering so, in colder climates, greenhouse conditions apply. Leucadendrons are readily propagated from seeds, or cuttings may be ordered from a specialist nursery.

## *Leviticum*

LOVAGE

*The distilled water of Lovage cleareth the sight and putteth away all spots, lentils, freckles and redness of the face, if they be often washed therewith . . .*

John Gerard, *The Herball,* 1597

***Apiaceae*** (Parsley family).

***Scent:*** Powerful, aromatic, thyme-like.

***Description:*** Lovage is an aromatic perennial plant, native to the Mediterranean, which was once highly valued for both its culinary and medicinal uses. Although it is not indigenous to England and Europe, it now grows wild there, having been extensively cultivated in the Middle Ages when

*Levisticum officinale*

Thomas Tusser advised it was one of the 'necessary herbs to grow in the garden of physic'.

The stems and leaves have a strong, celery-like flavor and may be used in salads, stews or casseroles. The new young stems may be candied, like angelica, and enjoyed as a sweetmeat or a decorative device. (NOTE: lovage is quite strong, so only use a small piece when cooking.) Lovage tea and lovage cordial—made by steeping the seeds in brandy—were beverages popular for their 'comforting and warming' effects. The seeds were also mixed in cake batter or used when preparing sauces for game. Parkinson said:

*The whole plant and every part of it smelleth strongly and aromatically and of a hot, sharpe, biting taste. The Germans and other Nations in times past used both the roote and seede instead of Pepper to season their meates and brothes and found them . . . comfortable . . .*

Lovage was also used as a prophylactic against cholera and in the healing of ulcers. The distilled water of lovage and the plant's juice have been tried as a method of bleaching freckles and also in baths for their deodorant ability, with Thomas Hyll succinctly noting in *The Gardener's Labyrinth* (1577) that '. . . this herbe for hys sweete

### LOVAGE AND MUSTARD SEED PICKLE

This goes well with cold ham, bacon, chicken or vegetable dishes. Or, try serving it with soft white cheese and plain cracker biscuits.

❖

*3 lemons*
*salt*
*apple cider vinegar*
*1–1½ tbspns chopped lovage*
*2 tbspns mustard seed*
*1 tspn Dijon mustard*
*3 onions, diced*
*2oz/60g sultanas*
*cinnamon stick*
*allspice and sugar, to taste*

*Chop lemons into tiny chunks and remove pips (seeds). Place in a mixing bowl and sprinkle liberally with salt. Cover with vinegar and let stand for 24 hours. Place lemons in a saucepan and add remaining ingredients. Bring to the boil, then cover and simmer until lemon pieces are well softened and pulpy, and the liquid is reduced and syrupy. Remove cinnamon stick. Spoon pickle into sterile jars and cap securely.*

savoure is used in bathe . . .' A lovage bath is very relaxing and refreshing. Bunches of dried lovage hung up in the kitchen will give a sharp, fresh scent to the room.

***Species:*** *L. officinale* (yellow flowers; 5ft/1.5m).

***Cultivation:*** Lovage grows readily from seed in most soil types. Sow in late summer, water well the following summer and cut back in autumn after flowering. According to Thomas Hyll, lovage '. . . joyeth to growe by wayes and under the eaves of a house, it prospers in shadowy places and loves running water', so it is ideal for overly moist spots around the house. Propagate plants by root division in spring time.

### LOVAGE AND JUNIPER BATH VINEGAR

Lovage has long been used as a natural deodorant—in 1577, Thomas Hyll wrote '. . . this herbe for hys swete savoure is used in bathe'. The cider vinegar in this recipe will help to soften skin and is suitable for those troubled with pimply blemishes or acne as it balances the skin's acid mantle and refines large pores.

*3 tbspns dried lovage*
*9fl oz/250ml apple cider vinegar*
*1½fl oz/50ml witch hazel*
*5 drops oil of juniper*

*Mix all the ingredients together in a lidded jar and steep on a sunny window sill for 2 weeks, shaking thoroughly each day. Strain and apply to affected area with a moistened cottonwool ball or add 3–4 tablespoons to a bath.*

## *Lilium*

LILY

*The Lily is an herbe with a white flower; and though the leaves of the flowre be white yet within shineth the likenesse of gold, the Lily is next to the Rose in worthinesse and noblenesse . . . Nothing is more gracious than the Lily in fairness of colour, in sweetnesse of smell and in effect of working and vertue . . .*

Bartholomaeus Anglicus, medieval scholar

*Lilium longiflorum*

***Liliaceae*** (Lily family).

***Scent:*** Orangey, spicy or honeyed, depending upon species.

***Description:*** This large genus of flowering bulbs is one of the oldest cultivated plants and is steeped in history. The ancients believed the dazzling white Madonna Lily, *Lilium candidum,* came forth from Juno's breast milk, and they consecrated the flower to her accordingly. Lilies have always symbolized purity and were used by religious artists and poets to typify all that was good and beautiful. Chaucer, in *The Second Nun's Tale,* gave this charming explanation of the origin of St Cecilia's name, 'Heaven's Lily':

*First wol I you the name of Sainte Cecilie*
*Expoune as men may in hire stories see:*
*It is to sayn in English, Heavens lilie,*
*For pure chasteness of virginitee,*
*Or for the whitnesse had of honestee,*
*And grene of conscience, and of good fame,*
*The swote savour, lilie was her name . . .*

Lilies are thought to have reached England and Europe from the East during the time of the Crusades. They became a familiar plant during Elizabethan times and Gerard referred to 'Our English White Lily [which] groweth in most gardens in England.' The flowers and roots and leaves were used for healing wounds. The roots were also used for removing freckles and lightening the skin. An old recipe gives instructions for boiling the roots in water until the water was reduced by a third, then applying this concoction to the face on nine successive nights. The roots were also roasted and mixed with oil of roses to 'take out the wrinkles on the face'. Although widely used for their elegant flowers and scent, lilies were thought an unlucky flower to bring indoors, as they were regarded as harbingers of illness, cold and death.

***Species:*** Of the 90 or so species of lilies available today, only about a third are scented. Key names to look for are *Lilium auratum,* the Lily of Japan (white flowers splashed with gold and crimson; 6–8ft/1.8–2.4m; late summer flowering); the Madonna Lily, *L. candidum* (white trumpet-shaped flowers; 4–5ft/1.2–1.5m; early summer flowering); the extremely exotic-looking *L. henryi,* discovered in the late nineteenth century in the Yangtze River region (apricot flowers on glossy arching stems; 8ft/2.4m) and the Bermuda or Easter Lily *L. longiflorum* (white flowers with golden anthers; 2ft/60cm), which is cultivated commercially for use as a cut flower throughout the world.

***Cultivation:*** Some species of lily are lime tolerant, others are not; check with your nursery specialist as to what sort of soil the different varieties prefer. Most Eastern-derived forms: *L. bakerianum, L. brownii, L. candidum* and *L. cernum,* for example, prefer a gritty, well-drained alkaline soil in an open situation with plenty of top dressing. The European species, such as *L. auratum,* will grow best in an acid soil in a shady spot. Not all lilies are pleasantly scented; avoid *L. pyrenaicum,* of which Gertrude Jekyll derisively said: 'May dogs devour its bulbs'!

## Limnanthes

POACHED EGG PLANT

***Limnanthaceae*** (Limnanthes family).

***Scent:*** Gentle and sweet.

***Description:*** These tiny annuals, which grow easily as edging plants, are native to America, particularly California. Though regarded as weeds in California, their pretty white cuplike flowers (about 1in/2.5cm across) and warm, honeyed scent show it is important not to overlook wildlings as a potential storehouse of fragrance for your garden. Limnanthes are minute treasures, growing only to about 6in/15cm tall. With their carpeting habit, they are good plants for setting in cracks between crazy paving stones or chinks in a retaining wall. Limnanthes are much loved by bees and butterflies and will attract them to a garden. The fragrance of the flowers varies according to the temperature, and is strongest on a warm, still, summer's day, intensifying still further toward evening.

*Limnanthes douglasii*

***Species:*** *Limnanthes douglasii* (yellow and white flowers reminiscent of poached eggs, hence their common name; 6in/15cm). The species was named after Scottish botanist James Douglas who cataloged the seed in the early nineteenth century and introduced it to England.

***Cultivation:*** Limnanthes are a hardy, frost tolerant annual. Sow seed in autumn to flower from early spring right through summer. Plant in a sunny place; Limnanthes will self-sow very easily and pop up in the most unexpected places.

### MY FAVORITE POTPOURRI

This has a very pretty scent, soft and relaxing. Try it in a guest bedroom or in a beautiful china jar near the front door.

*3½oz/100g chamomile daisies*
*3½oz/100g pink rose petals*
*2oz/50g pink carnation flowers*
*1oz/25g lemon balm*
*1oz/25g meadowsweet*
*1 crushed vanilla pod*
*1 tspn allspice powder*
*1 tspn cloves, crushed*
*1 tspn coriander seeds, crushed*
*2 tbspns orris root, powdered*
*essential oils—rose and carnation*

## Lonicera

HONEYSUCKLE

*Sleep thou and I will wind thee in my arms, . . . so doth the woodbine, the sweet honeysuckle gently entwist . . . Oh, how I love thee! How I dote on thee!*

William Shakespeare, *A Midsummer Night's Dream*

***Caprifoliaceae*** (Honeysuckle family).

***Scent:*** Sweet, honeyed, penetrating.

***Description:*** This group of deciduous climbing or standard shrubs, native to the Northern Hemisphere, is famed for its extremely fragrant flowers. These flowers are tubular and nectar is secreted at the base of the tubes, which attract

*Lonicera japonica*

## EXTRA-RICH HONEYSUCKLE OIL

This recipe works wonders on rough or chapped hands, and will help heal any nicks, split cuticles or bruises. Gardeners take note! Apply to your hands and then put on an old pair of cotton gloves before putting on heavy gardening gloves.

❖

*2 tbspns honeysuckle oil (see method)*
*1 tspn raw honey*
*2 tbspns sesame oil*
*2 tbspns almond oil*
*1 tbspn combined glycerine and rosewater*

*To make honeysuckle oil, infuse flowers in almond or avocado oil for 4–6 weeks in a sunny spot, turning jar regularly; strain. Melt honey in a double saucepan, then whisk together with the oils, glycerine and rosewater. Pour into sterile jars and cap securely.*

night-flying moths to effect cross-pollination. The climbers are more popular and showy than the shrub form of *Lonicera*. They have long been grown in English gardens, particularly during Elizabethan times when they were a favorite for covering arbors and bowers. Shakespeare refers to this in *Much Ado About Nothing* when Margaret is bid:

> *. . . run thee to the parlour;*
> *where shalt thou find my cousin Beatrice*
> *Proposing with the prince and Claudio:*
> *Whisper in her ear and tell her, I and Ursula*
> *Walk in the orchard, and our whole discourse*
> *Is all of her; say that thou overheards't us,*
> *And bid her steal into the pleached bower,*
> *Where honeysuckles, ripened by the sun,*
> *Forbid the sun to enter . . .*

Country names for honeysuckle included 'woodbine' and also 'love bind', because of its tightly embracing habit. It was an emblem of love and affection and Milton wrote, in *The Flower and The Leaf*, that '. . . those that wear chaplets on their head of fresh woodbine, be such as never were to love untrue, in word, thought, nor deed, but are steadfast.'

For best effect, plant honeysuckle by a window that will be open to the air on summer evenings because the perfume of the flowers becomes more pronounced as night falls. To achieve the traditional look of the cottage garden, allow honeysuckle to clamber up and down trees and festoon hedges with its fragrant flowers.

Honeysuckle flowers may be harvested and used in many ways. They can be macerated with a good quality odorless oil, such as almond, which is suitable as a massage oil. They are also a charming cut flower and a dainty choice for an old-fashioned 'tussie-mussie', or posy of sweet-smelling flowers. Arrange sprigs of honeysuckle, clove pinks and forget-me-nots around a large central red rose or pink camellia and offer as a pretty gift. And, to recapture the slower, drowsy days of a childhood summer, suck the nectar from the base of the flowers.

***Species:*** *Lonicera periclymenum*, the Common Honeysuckle (creamy yellow flowers with reddish outer sides; summer flowering); *L. japonica* (red and white flowers; autumn flowering) and *L. caprifolium* (yellow and purple flowers; midsummer flowering). Some of the shrubby honeysuckles form good hedging and several are winter-blooming—look for *L.* x. *purpusii* and *L. fragrantissima,* which both have white flowers. A very attractive variety is the hybrid *L.* x. *brownii* with its lush glowing red flowers, but it is scentless.

***Cultivation:*** Honeysuckle is hardy and grows easily in all soil types, either in full sun or semi-shade. In fact, the problem is that usually a plant can become invasive and take over other plants. Regular pruning and thinning out old stems each autumn will help keep honeysuckle under control.

## *Luculia*

***Rubiaceae*** (Madder family).

***Scent:*** Soft, incense-like.

***Description:*** A group of very beautiful winter-flowering large shrubs with broad leaves that change color as the season gets cooler, and sprays of pink or white flowers, which are very fragrant. Luculias are native to Southeast Asia and were introduced to England from the mountainous ranges of Bhutan.

*Luculia gratissima*

***Species:*** *Luculia grandifolia* (white flowers), *L. gratissima* (medium pink flowers) and *L. pinceana* (very pale pink flowers) all grow to at least 5ft/1.5m; if they are happy in their situation, they can grow much higher than this.

***Cultivation:*** Luculias require good drainage; they hate having 'wet feet'. Plenty of sun is required for the plant to flower and it needs to be fed with some rich compost. They are half hardy and will do well in most climatic conditions, though they need to be wintered under glass in areas subject to frost.

## *Lupinus*

LUPIN

***Fabaceae*** (Pea family).

***Scent:*** Perennial forms smell peppery or clove-like; annual forms are sweeter, with overtones of violet or vanilla.

***Description:*** With their elegant spires of pea-like flowers that have the sweet, mossy scent of the laburnum family, lupins have long been a favorite in cottage gardens. They are thought to have come originally from Egypt and the Greeks and Romans used them both as food and medicine, with Pliny saying there was '. . . nothing more wholesome and lighter of digestion than white lupins (*Lupinus albus*), eaten dry.' The name 'lupin' comes from the Greek 'lupe', meaning grief, referring to the extreme bitterness of the seeds if they are not boiled before being eaten. Powdered lupin flowers were mixed with lemon juice, goat gall or both and used as a face mask; in fact, Culpeper strongly recommended this treatment for removing smallpox scars, which were so widespread at the time.

The original lupins were annuals and were probably either white (*L. albus*) or yellow (*L. luteus*). From these have evolved the many brightly colored varieties found today: golden, pink, mauve and blue. A relatively recent introduction was the Perennial Tree Lupin (*L. arboreus*), from California in 1793.

*Magnoli lennei*

***Species:*** Annual forms: *Lupinus luteus,* or Spanish Violet (yellow flowers with violet scent); *L. albus* (white flowers); and *L. hartwegii* (blue flowers; unscented). All are approximately 2ft/60cm high. Perennial forms: *L. arboreus* (racemes of profuse golden flowers; clover scented), which grows to about 5ft/1.5m tall.

***Cultivation:*** Lupins require a well-drained soil and a little neglect. They do not grow well in a rich, composted soil and actually prefer a thin, sandy one. The seeds should be planted in spring for early summer flowering through to the beginning of autumn. Place them up the back of the flower border as they can start to look a little raggedy.

## *Magnolia*

*Faint was the air with odorous breath of magnolia blossoms, spread high as a cloud, high over head!*

William Wordsworth, *Ruth, or Influences of Nature*

***Magnoliaceae*** (Magnolia family).

***Scent:*** Rich and fruity.

***Description:*** These evergreen or deciduous flowering trees and shrubs have long impressed gardening writers with their beauty. In 1895, Donald McDonald wrote in *Scented Flowers and Fragrant Leaves* that the magnolia combined the '. . . beauty of the rose and the odour of the lily'.

Magnolias are widely represented throughout Asia, Europe and America. The Chinese were particularly fond of magnolias, especially the *yulan*, or *Magnolia denudata*, which always featured on their porcelain and ornamental screens. They had a charming folk tale to explain this tree's habit of producing its ivory and purple violet-scented flowers before it bore its leaves:

> *They say that the owl, in its wisdom, comes every spring when the magnolia's buds are just about to unfurl and says: 'Hold fast! Hold fast! If you speak now you will lose your influence for the whole year!' But the magnolia, being proud and vain, does not listen to the warning and opens its petals so its perfume 'speaks' out in the air—and it has no perfume left for the rest of the year.*

***Species:*** The evergreen *Magnolia grandiflora* (creamy, lemon-scented flowers; 100ft/30m; summer flowering) is an impressive spectacle in full flower. Henry Phillips wrote that the flowers '. . . perfume the air for a considerable distance around with the most agreeable odour which at one moment reminds us of the jasmine or lily of the valley, the next of the violet mixed with the apricot.' Other stunning summer-flowering magnolias are *M. wilsonii* (white flowers with crimson anthers) and the dwarf Japanese *M. sieboldii* (syn. *M. parviflora*), which has dainty cup-shaped pink and white flowers and only grows to 4ft/1.2meters. Deciduous types, which tend to flower throughout spring, are: *M. denudata* (white and purple flowers, violet-scented); *M. stellata*, the Star Magnolia (white star-shaped flowers, very daintily formed) and the fruitily scented Willow Leaf Magnolia from Japan, *M. salicifolia* (white starry flowers; 20ft/6m). Also deservedly popular is the exquisite night-scented Port Wine Magnolia, known as *Michelia figo* (*Michelia* being a closely related group).

***Cultivation:*** Magnolias grow best in an acid soil and full sun. They should be protected from the wind and, preferably, be planted in an open situation rather than cramped in a flower bed, so they may develop their graceful form. In my neighborhood, there is a house where two *M. stellata* have been planted on either side of the front gate and allowed to grow together, so visitors are greeted with a beautiful show and unexpected fragrance.

## *Mahonia*

OREGON GRAPE

***Berberidaceae*** (Barbary family).

***Scent:*** Extremely sweet; elusive; rather like lily of the valley.

*Mahonia aquifolium*

***Description:*** The mahonias are handsome, winter-flowering, quick-growing evergreen shrubs, native to China, Japan and North America. The leaves are like holly and the golden flowers are borne in scented spikes. You must come close to enjoy the fragrance because it is so subtle. The flowers are followed by extremely decorative blue-bloomed fruits, explaining the derivation of its name, Oregon Holly Grape (it is also the floral emblem for this American state). The root extract has long been used by herbalists as a tonic and alterative. Combined with cascara, Oregon Grape is used in many patent preparations to relieve constipation. A mahonia will also provide much needed winter color in a shrub border or shady spot, but it should be placed where you will often pass by so the elusive scent is not wasted.

***Species:*** *Mahonia japonica*, native to Japan, is deliciously perfumed (clusters of yellow flowers; 4–5ft/1.2–1.5m). Other choices are *M. aquifolium*, the Oregon Holly Grape (racemes of golden flowers; 3ft/1m); *M. bealei* (erect spikes of golden flowers; 10ft/3m) or the very attractive *M. fremonti* (golden flowers; 7ft/2m), which has gray leaves to offset its sunny colored flowers.

***Cultivation:*** All the mahonias are extremely hardy and tough as well as being colorful and decorative. They grow well in most soils and will tolerate shade and frost quite well. Mahonias should be planted in spring and may be propagated from cuttings taken when the shrub is pruned each summer.

## *Malus*

CRAB APPLE

***Rosaceae*** (Rose family).

***Scent:*** Sweet and pronounced.

***Description:*** Crab apples are extremely pretty deciduous trees with superb fragrant blossoms in spring. The pink-stained buds and cupped white flowers are an entrancing sight in late spring, and are later followed by little amber and red apples. These fruits are edible and were the forerunner of today's orchard apple. Crab apples were often seen on the medieval menu, being used in syrups, conserves and puddings. Shakespeare referred to 'roasted crabs', which were served around the Yuletide hearth with mulled wine or ale.

***Species:*** *Malus purpurea* is one of the most outstanding forms (wine-red flowers and reddish copper foliage; 20–25ft/6–8m) but it is, unfortunately, almost devoid of scent. The following bear lightly scented blossoms: *Malus angustifolia* (apricot/pink flowers; violet-scented); the Wild

*Malus* x *purpurea*

Crab Apple, *M. sylvestris* (pink and white flowers; violet scented; 20ft/6m), and *M. hupehensis* (white flowers; 15ft/4.5m). Although the blossoms are very fragile and disappear with the first breeze, crab apples are extremely decorative with an attractive upright branching habit and they are extremely lovely in small gardens.

***Cultivation:*** Much like the orchard fruits, the decorative crab apples appreciate plenty of sunshine and a rich, well-drained soil. Some varieties are self-pollinating but others need to be planted by a pollinating variety in order to fruit, so check with your nursery specialist before planting.

## *Matthiola*

STOCKS

*. . . but the few lingering scents*
*of streaked pea, and gillyflower, and stocks*
*of courtly purple and aromatic phlox . . .*

Robert Bridges

***Brassicaceae*** (Mustard family).

***Scent:*** Exotic, clove-like.

***Description:*** This group of annuals and biennials was named for the great sixteenth century

*Matthiola incana*

Italian botanist Mattioli. Stocks were first commonly described as 'stock gillofers' or 'stock gelouers', because their strong, sweet scent reminded early herbalists of carnations and pinks. John Gerard said they were '. . . greatly esteemed for the beautie of their flowers and pleasant sweet perfume' and the poet Thompson in *The Seasons* described the '. . . lavish Stock, that scents the garden round'. Perhaps it was Francis Bacon, the lively seventeenth century garden essayist who best put his finger on the nature of the elusive and elegant scent of stock. He described it as '. . . far sweeter, in the air (where it comes and goes like the warbling of music) than in the hand.' Indeed they are more useful in a scented garden than for use as a cut flower indoors. Tuck them under windows and plant them with more dramatic though unscented bedding plants.

Stocks were once thought to have powerful love associations and Charles II kept pots of night-scented stock in his chambers for the express purpose of creating a seductive atmosphere. Perhaps this was why the rakish Lord Byron insisted they be grown in every available place around his home, Newstead Abbey, in Nottinghamshire. On a more prosaic note, the leaves of stock may be used as an interesting, somewhat peppery, accompaniment for salads and stews. Old almanacs also yield recipes for syrup and cordials made by distilling the scented flower petals.

***Species:*** *Matthiola bicornis*, the Night Scented Stock (whitish/lilac flowers; 20in/50cm) is the original form referred to by Gerard and Turner. The little dingy flowers are closed and drooping during the day, only opening at night. Lavender and purple forms are also available. Despite the unprepossessing flowers, Night Scented Stock is a storehouse of intoxicating scent, which intensifies in a still, twilit garden. In spring, the fast-growing *M. incana* has light purple flowers borne in spikes (3-6in/7-15cm long).

***Cultivation:*** Sow seed in spring directly into the garden, as stocks do not like being disturbed once they are established. The group known as Brompton stocks are spring-flowering and fast growing and may be treated as biennials with the seed being sown in summer and seedlings set in autumn to bloom the following year. Stocks are attractive as a garden or path edging, or about a rock garden or border; they will tolerate some shade.

## *Melaleuca*

CAJEPUT TREE; PAPERBARKS

***Myrtaceae*** (Myrtle family).

***Scent:*** Leaves are aromatic: sweet, lemony or antiseptic/tea-tree like, depending on species.

***Description:*** A group of evergreen trees or shrubs that is now native to Australia, having been originally introduced from Indonesia and Malaysia. Melaleucas bear long spikes of very soft and pretty fluffy white, cream or pink and mauve flowers. These flowers are fragrant in some species, but it is the leaves and twigs that give the plant its distinctive aroma. The flaky layers of white bark give it its common name of Paperbark Tree. The volatile oil, cajeput, is distilled from the fresh leaves of the species *Melaleuca leucadendron* and is used both as an inhalant preparation or, diluted with olive oil, as a treatment for skin infections. Cajeput oil has a very warming effect and this may be detected just by rubbing the hands vigorously with a few leaves. This will stimulate the circulation and produce a tingling effect—an old bush trick for coping with winter months in the outback.

***Species:*** *Melaleuca leucadendron* (white fluffy flowers; 20ft/6m); *M. squarrosa* (creamy, lemon-scented flowers; 10ft/3m); *M. thymifolia* (white and mauve flowers; 6½-30ft/2-10m).

***Cultivation:*** Melaleucas are good screening plants and may be planted along fence lines or paths, but not too close to the house as, with the exception of *M. squarrosa* (honey-scented flowers), they will grow to be very tall. Most are found in the swamplands of northern New South Wales or Queensland and will do well, even if drainage is poor. Only one species, *M. quinquenervia*, must have a well-drained sandy soil. Greenhouse cultivation is required in cold climates; otherwise, melaleucas are very easy to please.

### SCENTED WATER

You may like to try making this very simple and delicate scent using freshly picked fragrant blooms from your garden.

❖

*17½fl oz/500ml distilled water*
*2 tbspns lavender flowers*
*1 tbspn rose-scented pelargonium leaves*
*1 tbspn lemon verbena leaves*
*2 tspns grated orange zest*
*2 tbspns brandy*

*Pour water over flowers, leaves and zest. Cover and steep for 2 days. Chill, strain and stir brandy through the mixture. Store in an atomizer and keep in the refrigerator for a refreshing facial spray on a hot day.*

## *Mentha*

MINT

### To Make Syrup of Mint

*Take a quart of the syrup of Quinces before they are full ripe, juice of Mint two quarts, an ounce of Red Roses, steep them twenty-four hours in the juices, then boil it till it is half wasted, strain out the remainder and make it in to a syrup with double refined sugar.*

from *The Receipt Book of John Nott,* Cook to the Duke of Bolton, 1723

***Lamiaceae*** (Mint family).

***Scent:*** Crisp, cool, sweet, refreshing.

***Description:*** Mint takes its name from Menthe, the nymph beloved of Pluto, who was said to have been turned into the herb by Pluto's jealous wife, Proserpine. Mints came originally from southern Europe and the Mediterranean and were introduced to England during the Roman Conquest. Mrs Leyel writes that mints '. . . grow better and smell sweeter in England than they do anywhere else'. However, they are now naturalized widely throughout the world and, until recently, Michigan in Ohio was one of the largest cultivators of mint for commercial production and would certainly not have agreed with Mrs Leyel's parochialism.

Mints make an admirable ground cover in a scented garden, bestowing a soft cushiony effect and fresh scent so deftly described by Geoffrey Chaucer:

*Then went I forthe on righte honde*
*Down by a little path I fonde*
*Of mintes full and fennel green . . .*

There are many different species in this group of aromatic perennial herbs, from the common and garden cooking mint to the delicious and fragrant apple mint, spearmint, ginger mint and orange mint. In 1597, John Gerard wrote of these 'divers sorts of mints, some of the garden, others wilde or of the field; and also some of the water', and of their many applications. Though mints are strongly associated with the culinary arts, and are used to make mint sauce, as a flavoring for cups or cocktails and to enhance salads or spring vegetables, just to name a few, Culpeper also recommended a poultice of hot rose petals and mint leaves to cure insomnia. Mints were also one of the herbs strewn in churches and in homes to keep mice away. Peppermint tea is a time-honored remedy for indigestion and nausea.

Mint was once used to keep milk from curdling and *The Good Housewife's Handmaid* (1588) tells us that:

*Mintes put into milke, it neyther suffereth the same to crude, nor to become thicke, insomuch that layed in curded milk, this would bring the same thinne againe . . .*

Interestingly, herbalists will still prescribe mint tea for people with a milk intolerance, for it helps them digest dairy products. Things have not changed very much, in fact, from when ninth-century monk Walafrid Strabo, wrote:

### PEPPERMINT TRAVEL PILLOW

Many people become apprehensive when traveling, whether through fear or actual motion sickness. A gift much appreciated by the queasy traveler is a soft neck-cushion—kidney shaped is best, as it will support the head—filled with:

*2oz/50g peppermint*
*½oz/15g mint-scented pelargonium*
*2oz/50g lemon verbena*
*2oz/50g lavender*
*1 tbspn crushed lemon zest*
*1 tspn nutmeg chips, crushed*
*3 drops peppermint oil*

▸ The lushness of the primula's colors are complemented by the delicate rosemary.

*They say that Eastern doctors will pay as much for it as we pay here for a load of Indian pepper . . . Believe me, my friend, if you cook some pennyroyal and use it as a potion or a poultice, it will cure a heavy stomach . . . when the sun is blazing down on you in the open, to prevent the heat from harming your head, put a sprig of pennyroyal behind your ear.*

***Species:*** *Mentha piperita* or peppermint is the variety cultivated commercially for its aromatic oil used in cosmetics. This type of mint is grown extensively in England, especially Mitcham, for the damp climate improves the flavor and fragrance content of the leaves. Also well known are *M. spicata* or spearmint (green smooth leaves); *M. piperita* var. *citrata,* the Eau de Cologne or Bergamot Mint (large flat oval leaves); Apple Mint *M. suaveolens* (cream and green variegated leaves) and the Wild Water Mint, *M. aquatica.* Pennyroyal, *M. pulegium,* is one of the prettiest ground covers to have in a scented garden. It is dense and sturdy with dainty little mauve flower spikes in summer. It can be used to great effect as a lawn, either by itself or mixed with sweeps of different mints. Another charming idea is to mold a 'seat' from earth in a shady spot in the garden and plant it with clumps of pennyroyal. Not only will such a pretty spot refresh the senses of sight and smell, it will also help repel flies.

### CRYSTALLIZED MINT LEAVES

❖

*mint leaves*
*2 egg whites, lightly whisked*
*caster (superfine) sugar*

Preheat oven to 250°F/120°C. Gently rinse leaves and pat dry with absorbent paper to remove moisture. Paint leaves on both sides with egg white. Dust with caster sugar and place on waxed paper in the oven for 10 minutes, until crisp.

❖ ❖ ❖

***Cultivation:*** Mints are hardy, quick-growing and will thrive in most soils. If they have a fault, it is that they are so vigorous they can be intrusive, spreading through a garden, popping up in flower boxes and cracks in the paving. They will grow in full sun but the flavor of the leaves is improved if there is a little shade during the day. Mints appreciate plenty of moisture and need to be well watered, especially in summertime. If harvesting the leaves for use in the kitchen, cut mint back hard during the summer to prevent it becoming straggly. Mints are easily propagated by taking root cuttings in autumn.

## *Mirabilis*

MARVEL OF PERU, FOUR O'CLOCK PLANT

*The marvell of the world comes next in view,*
*At home, but stil'd, the marvail of Peru.*
*(Boast not too much, proud soil, thy Mines of Gold,*
*Thy veins, much wealth, but more of poison hold)*
*Bring o'er the root, our colder Earth has power*
*In its full beauty, to produce the flower;*
*But yields for issue no prolifick seed,*
*And scorns in foreign lands to plant and breed.*

Abraham Crowley

*Mirabilis jalapa*

***Nyctaginaceae*** (Four O'Clock family).

***Scent:*** Fruity, sweet, refreshing.

***Description:*** This is an exotic group of tuberous-rooted plants that are wildlings of the tropics of South America. Seeds of the Marvel of Peru were introduced to Europe by Spanish Conquistadores returning from South America in the sixteenth century. The Peruvian women used the dye in the flower petals as a natural rouge, rubbing the petals on their cheeks and hands; the dried roots were employed as a vermifuge to expel worms from the body. The small petunia-shaped flowers come in a wide color range, including yellow, red, deep crimson and white and they flower continuously through the summer months. These flowers do not appear until late afternoon, when the sun moves off them and the air begins to cool; then they unfurl, releasing their rich and refreshing bouquet. The cherry-red kinds are the most fragrant and, in contrast to other night-scented plants whose fragrance can be quite heavy and even cloying, the Marvel of Peru's scent could be described as light and moving rapidly around the garden at twilight.

***Species:*** *Mirabilis jalapa*, or the Marvel of Peru (crimson, pink or white flowers; 2ft/60cm). The tiny variety 'Pygmea' is a charming choice for a potted houseplant, preferably in a bedroom.

***Cultivation:*** Although a perennial, mirabilis is tender and should be grown as an annual. It may be cultivated either from seed or, rather like dahlias, dig the tubers up in autumn and replant the following year in late spring. If the plant is in a sunny situation and does not need to be dug up each year, remember that the bush does grow quite quickly and will need some sort of support or the branches may get broken in heavy winter rains. Plenty of sunlight, moisture-retentive soil and a heavy hand with the fertilizer are all desirable for growing this exotic plant successfully. A clever idea is to place one where it will only receive half a day's sunshine; this way, its 'working hours' will be lengthened.

## *Monarda*

BERGAMOT

*Speak not—whisper not,*<br>
*Here bloweth thyme and bergamot,*<br>
*Softly on there every hour,*<br>
*Secret herbs their spices shower.*

Walter de la Mare

***Lamiaceae*** (Mint family).

***Scent:*** Delicious, orange-like.

***Description:*** A group of six or seven perennial species native to North America and Canada. During summer, bergamot produces shaggy flower heads in a wide variety of colors: pink, rose, purple and the most popular, red. The flowers are rich in nectar and particularly attractive to bees. They were named after the sixteenth century Spanish physician, Dr Nicholas Monardes, who first described them in his book *Joyfull Newes out of the newe found worlde*. The common name, bergamot, was given because the aromatic perfume of the leaves, stems and roots was thought to be

*Monarda didyma*

### NUTTY BERGAMOT NOG

This soothing and delicious beverage is a marvelous restorative for a patient with a sore throat, or for a feverish child.

❖

*9fl oz/250ml milk*
*2–3 tbspns bergamot, bruised*
*2 eggs, separated*
*1 tbspn ground almond meal*
*sugar, to taste*
*nutmeg or allspice*

*Heat milk to simmering point and infuse bergamot for 15 minutes; strain. Beat egg yolks, almond meal and sugar together and add to milk. Whisk egg whites until frothy and fold through. Sprinkle with a little of either spice and serve immediately.*

similar to the bergamot orange, grown at Bergamo in Italy. American pioneer women used bergamot to brew 'Oswego-tea', which they used to cure a multitude of ills. This tea eventually became extremely fashionable, being drunk in great quantities around the time of the famous Boston tea party. Henry Phillips, writing in his *Flora Historica* (1824) said that '. . . many persons prefer the infusion of the leaves to the tea of China.' It is still the key flavoring ingredient used in Earl Grey tea. Bergamot leaves and flowers may be used in salads, stuffings and drinks; dried and crumbled, the leaves add a refreshing tang to potpourri and sachet mixes.

***Species:*** *Monarda didyma* (whorls of red or purple flowers; 3ft/1m), also known as bee balm, has several cultivars worth seeking: 'Cambridge Scarlet' with handsome, bright red flowers, 'Blue Stocking', with deep violet flowers and 'Enfield Gem' with rosy mauve flowers. Wild bergamot, or *M. fistulosa,* was called the 'wild mint of America' by John Tradescant in 1637. It has lavender flowers and, occasionally, a white form is available, which is lovely in a garden planted for night-time enjoyment.

***Cultivation:*** Grow *M. didyma* in a moist semi-shaded position. It is easily raised from seed; seedlings should then be planted in the garden in clumps and fed well with compost or manure. When it has finished flowering, thin out the flower shoots that the plants have produced and replant in clumps to produce next year's harvest. *M. fistulosa* prefers a drier soil.

## *Murraya*

ORANGE JESSAMINE OR CHINESE BOX TREE

***Rutaceae*** (Rue family).

***Scent:*** Intense, jasmine-like.

***Description:*** This group of evergreen trees or shrubs is native to China, India, Indonesia and Polynesia. They are very attractive with dense, dark green glossy leaves and masses of fragrant, orange blossom-like flowers. These flowers are followed by red berries, which are also quite attractive. In

*Murraya paniculata*

some species the leaves have a powerful spicy smell and they are used for flavoring curries. Murrayas are great in full sun, either in a container or massed as a hedge. They have a natural, rounded habit of growth but will take firm clipping for a formal look, or even as a subject for topiary.

***Species:*** *Murraya paniculata* (white flowers; 20ft/6m).

***Cultivation:*** Murrayas are hardy in temperate zones, preferring full sun and a well-drained soil. They are also cultivated as greenhouse specimens in frost-prone areas, and require a minimum winter temperature of 65°F/18°C.

## *Muscari*

GRAPE HYACINTH

***Liliaceae*** (Lily family).

***Scent:*** Incense, or musk-like; honeyed.

***Description:*** These delightful, free-flowering little bulbs were mentioned in 1614 by Crispin Van de Passe in his *Hortus Florids.* Their name, *muscari,* is a reference to the rich, musky scent of the flowers; just a few stems will scent a whole room. John Parkinson was to call it 'Pearls of Spain' and, in an attempt to describe the plant's sweet perfume, he recorded that it was '. . . like unto stock when it was newe and hot'. John Gerard wrote that grape hyacinths were '. . . kept and maintained in gardens for the pleasant smell of their floures, but not for their beautie'. He must have had a fit of pique when maligning this plant, for not only does it have a lovely perfume, but the bright blue or white spikes of tiny blooms, perched on their stems just like clusters of grapes, are very pretty indeed. Ruskin was perhaps more complimentary when he described the grape hyacinth as being: '. . . as if a cluster of grapes and a hive of honey had been distilled and pressed together in one small boss of celled and beaded blue . . .'

Grape hyacinths are also hardy and a perfect choice for a sunny rockery or for naturalizing in short grass. Potting several bulbs in a window box will ensure the fragrance is wafted indoors when the windows are opened for spring. Like many bulbs, grape hyacinths look their best when massed or allowed to naturalize and at their worst ranged in skimpy formal little rows. Use them in front of the flower bed as an edging plant but I strongly recommend that you do so *generously.*

*Muscari armeniacum*

***Species:*** *Muscari armeniacum* (bright blue bell-like flowers; 9in/20cm, honey sweet perfume); the Italian Grape Hyacinth, *M. botryoides* (blue or white flowers; 6in/15cm); and the quaint Tassel Hyacinth, *M. comosum* 'Monstrosum', which Gerard called 'the fair-haired hyacinth', referring to its purple and gold coloring.

***Cultivation:*** Muscari are very free flowering and quite tough. They will do equally well in poor dry soil beneath a tree or in good quality potting mix in a window box. They actually seem to do better in the former, provided there is plenty of sun.

NURSERY POTPOURRI

A soft and refreshing perfume for a baby's room.

❖

*9oz/250g rose petals*
*4½oz/125g lavender flowers*
*4½oz/125g lemon verbena*
*2oz/50g marjoram*
*2 tbspns powdered allspice*
*1 tbspn dried lemon zest*
*2–3 tbspns orris root powder*
*essential oils—rose or lavender*

*Mix all the ingredients together well in a large china bowl; add extra oil as desired.*

## *Myosotis*

FORGET-ME-NOT

*Forget me nots, all bathed in pearly dew,*
*Have opened wide their bonny eyes of blue;*
*And meadowsweet, like a young joyous bride,*
*Throws its light shadow in the crystal tide.*

Matthew Harman, *To Musella*

***Boraginaceae***

***Scent:*** Fresh, delicate; more pronounced in the evening.

***Description:*** Forget-me-not's botanical name, myosotis, means 'mouse ear' because the small fuzzy leaves curl like a mouse's ear. The forget-me-not is a much loved perennial with a romantic, if tragic, legend accorded to its sentimental name.

*Myosotis sylvatica*

Forget-me-nots are native to Germany and France, and it is said that a couple were strolling along the banks of the Danube on the evening of their betrothal when the girl saw the little blue flowers growing close to the water and, pettishly, demanded that her lover pick them for her. He attempted to do so, but fell in and drowned, crying out as he was swept away 'Lover! Forget me not!'

Forget-me-nots were sometimes incorporated on stationery as lovers' mottoes, poked into bouquets or embroidered on handkerchiefs given by women to their seafaring sweethearts. They were also chosen by Henry of Lancaster as his personal emblem in the hope that he, too, would not be forgotten by history.

A bed or 'carpet' of forget-me-nots is a very dainty sight. With their sky-blue faces and bright yellow 'eyes' one can appreciate Coleridge's disappointment when he could not find:

*. . . by rivulet or spring, or wet roadside,*
*That blue and bright eyed floweret of the brook,*
*Hope's gentle gem, the sweet Forget-Me-Not.*

Forget-me-nots are a long-lasting choice for a potpourri or dried flower arrangement and add color while other petals might look faded.

***Species:*** *Myosotis sylvatica* (blue/yellow flowers; 8in/20cm) is the main species. Lesser known ones include *M. alpestris* and *M. scorpioides*. The cultivars widely found in nurseries are usually hybrids of these. In addition to the deep blue forms, pink forget-me-nots are also available.

***Cultivation:*** Forget-me-nots are a very useful spring bedding plant, either growing beneath other plants or in a mixed border with tulips or daffodils. They are also pretty in pots. Forget-me-nots may be treated as biennials, with seed being sown in late spring as well as in autumn. Water sparingly. If left alone, forget-me-nots will rampage through a garden. During my first year in a cold climate city, I was ecstatic that, for the first time in my life, I could pick enough forget-me-nots to fill every vase in the house; the neighbors were very puzzled by my affection for this 'weed'.

## *Myrrhis*

SWEET CICELY, SWEET FERN, BRITISH MYRRH

*Sweet Chervil or Sweet Cis is so like in taste unto Anis seede that it much delighteth the taste among other herbs in a sallet.*

John Parkinson, *Paradisus in Sole Terrestris,* 1629

***Apiaceae*** (Parsley family).

***Scent:*** Sweet, aromatic, like myrrh or incense.

***Description:*** Sweet Cicely is a beautiful perennial plant and a very old denizen of country style gardens. It is native to the mountainous areas of northern Europe and the United States where it was used by the American Indians as a food. It was also one of the holy herbs used by Moses to make the sacred oil of the Tabernacle. Like several of the valued medicinal and culinary plants, Sweet Cicely was thought to have been blessed by the Virgin Mary, inspiring Vita Sackville-West to write:

*Tansy, thyme, Sweet Cicely,*
*Saffron, balm and rosemary,*
*That since the Virgin threw her cloak*
*Across it, so say cottage folk,*
*Has changed its flowers from white to blue . . .*

### SCENTED OVEN MITTS

These are very easy to make. Simply unpick a seam and tuck this mixture into the padding that covers your palm.

*2oz/50g lemon verbena*
*1oz/25g bay leaves, crumbled*
*1oz/25g rosemary*
*½oz/15g lovage*
*1 tspn cloves, crushed*
*1 tspn cinnamon*
*1–2 tbspns orris root powder*

*Then, sew it up. A fresh burst of fragrance will be released each time the mitt is squeezed around a hot dish.*

The feathery leaves that turn purple in autumn are delicious in salads or with fruit and taste very sweet and quite spicy, so explaining the derivation of one of the plant's common names. John Evelyn in his *Acetaria: A Discourse of Sallets* (1699) was emphatic that chervil should '. . . never be wanting in sallets as long as they may be had, being exceedingly wholesome and cheering the spirits.' He added that '. . . the rootes boiled and eaten colde are much commended for aged persons.' The roots are also quite sweet, and may be grated, raw, into salads or stir-fried in oil and vinegar as a refreshing side dish with poultry and game. The seeds are extremely aromatic with an attractive myrrh-like fragrance and taste like licorice drops when eaten. Around the house, the dried leaves may be used in potpourri mixes or placed in the linen press. The seeds, crushed in a mortar and pestle, should be rubbed over wooden furniture for a high gloss and lingering fragrance. The dried, lacy leaves are a pretty subject for a home-made pressed flower arrangement.

***Species:*** *Myrrhis odorata* (strongly scented bright green leaves and small umbels of white flowers followed by shiny black seeds; 5ft/1.5m).

*Myrrhis odorata*

SWEET CICELY AND ALMOND BUTTER

This is particularly delightful on scones, fruit buns, sweet biscuits or pancakes.

❖

*2 tbspns chopped sweet cicely*
*4½oz/125g unsalted butter*
*2 tbspns ground almond meal*
*1 tspn caster (superfine) sugar*
*2–3 drops almond essence (extract), to taste*

*Blend all the ingredients until the mixture is smooth and fluffy. Spoon into a small glass dish and refrigerate until served.*

❖ ❖ ❖

***Cultivation:*** Sweet Cicely is relatively slow-growing, and takes several years to attain its full height. However, the leaves may be pinched off and used when the plants are only a few inches high. Sweet Cicely prefers a light, well-drained soil and semi shade. Sow seeds in spring, or divide roots of existing plants in spring or autumn to propagate. Well-established plants self-sow readily, producing circlets of seedlings around them, which may be easily transplanted. Remember the plants may die back completely in winter, but they will reappear the following spring.

## *Myrtle*

*From the delightful perfume of the myrtle, the delicacy of its blossoms, and the gloss green of its perpetual foliage; it seems destined to ornament the forehead of beauty.*

Helen Milman, *My Kalendar of Country Delights*, 1903

***Myrtaceae*** (Myrtle family).

***Scent:*** Distinctive, fresh, slightly antiseptic.

***Description:*** Myrtles were sacred to Venus, the goddess of love, so it is not surprising to find they were used as a symbol of love. The goddess herself is depicted wearing a chaplet of myrtle and vervain and a sprig of myrtle was essential in wreaths worn at weddings by both the bride and groom and their guests. These wreaths became so famous throughout the Mediterranean that the flower market where they were sold became known as the 'myrtle market'. In Germany, potted myrtles are a traditional wedding gift and should be placed on either side of the door to repel evil influences.

All parts of the myrtle are fragrant. The leaves have a clean, refreshing scent and contain myrtol, which gives them natural disinfectant properties. Distillations of the leaves were used as a medicine and also as toilet waters. The flowers are aromatic and also edible, making a zesty garnish for a summer salad or cheese platter.

### MYRTLE AND YOGHURT CLEANSER

This recipe is particularly effective for adolescents and those with oily skin. Myrtle water was, after all, said to have been Venus' beauty secret.

❖

*5fl oz/150ml distilled water, boiling*
*2 tbspns myrtle flowers*
*5½oz/150g natural yoghurt*

*Pour the boiling water over the flowers and infuse; cool and strain. Blend myrtle water with yoghurt to form a smooth, runny lotion. Massage gently into skin before rinsing off with cool water and patting dry.*

*Myrtus communis*

***Species:*** *Myrtus communis* (white flowers; autumnal berries; 10ft/3-4m) is a small-leafed evergreen, which is a fine potted specimen. Myrtles lend themselves well to topiary and may be clipped into imaginative, and fragrant, shapes then taken indoors to dining or reception rooms. Also attractive is the Chinese Myrtle, *M. luma* (white flowers; 30ft/10m), which has richly terracotta-colored bark—striking against the snowy blossoms.

***Cultivation:*** Myrtles are easy to grow and satisfyingly quick to spring up; this makes them a good choice for hedging or masking an unattractive landmark. They do not tolerate frost; wheeling a potted specimen into a greenhouse is the solution for a myrtle in a cold climate.

## *Narcissus*

***Amaryllidaceae*** (Amaryllis family).

***Scent:*** Delicate, sweet, fresh.

***Description:*** This very large group, which includes daffodils, narcissi, jonquils and the Lent Lily, is perhaps the most widely grown of all bulbous plants. They are ideal for naturalizing, spring bedding or for use as a cut flower. The ancient Egyptians used narcissus flowers for their funeral decorations and, even after many centuries, the preserved remains may be recognized. Pliny maintains the narcissus was so named because of the supposed narcotic quality of its perfume and Socrates named it the 'Chaplet of the Infernal Gods' because of the bulb's poisonous effects. The more popular version of how these plants obtained their name comes from Ovid. According to him, Narcissus, the very handsome son of Cepnisus, caught sight of his reflection and, mistaking it for an exquisite water nymph, fell in love with it. He jumped in and drowned and when the gods tried to take up the body so they might give it funeral honors, they found only the lovely sweet narcissus, which was henceforth an emblem of vanity.

***Species:*** *Narcissus bulbocodium,* the little Hoop Petticoat Daffodil (yellow funnel shaped flowers; 6in/15cm) and *N. cyclamineus,* which has reflexed perianths similar to those of a cyclamen

*Narcissus* 'La Argentina' and *Narcissus bulbicodium* surrounded by *Muscari armeniacum.*

(4–8in/10–20cm). They require plenty of moisture and are ideal for naturalizing in damp, shady or otherwise difficult places in the garden. *N. jonquilla* (short cupped, golden flowers; 8–12in/20–30cm) has probably the sweetest scent of any of the jonquils, though some find it almost overpowering. An essential oil is obtained from the flowers and used in heavy types of perfumery, such as oriental or floral types. The Rush-Leaf Daffodil, *N. assoanus,* so named for its deep green, rush-like leaves, is also lovely, as is the Poet's Narcissus, *N. poeticus,* which has starry white flowers. The old Tazetta type of narcissi (*N. tazetta*) have clusters of flowers that are strongly perfumed; these are popular for forcing or for use as a florist's cut flower. A pretty centerpiece for the dining room table would be the contrasting varieties 'Paper White' and the golden 'Soleil D'Or', grown in glass forçing bowls or on pebbles standing in a dish of water. Some of the miniature types have a strong, cool, 'woodland' type of perfume: *N. canaliculatus* (orange flowers on bluish-green stems; 6in/15cm); *N. asturiensis* (golden fringed flowers; 4in/10cm) and the early flowering Lent Lily, *N. pseudonarcissus.* Wordsworth's famous *Ode to Daffodils* referred to the Lent Lily and it is still grown widely in meadows in the picturesque Lake District in England. *N.* x. *odorus*, the Spanish Campernelle, looks lovely nodding away in window boxes, and the clusters of flowers emit a delicious perfume.

***Cultivation:*** Narcissi require a long season of growth, so they should be planted out early, preferably in late summer, either in full sun or partial shade. Remember, they are happiest in a woodland glade, so bear this in mind when looking about for a suitable aspect. (NOTE: if you live in a very hot area, you will find it difficult to grow narcissi successfully.) In very damp soil, the bulbs should be set out on a layer of sand to avoid waterlogging. The smaller kinds are ideal for planting in rockeries, pots or window boxes, or for naturalizing in short grass. The little ones should be planted 2–3in/5–7cm deep, the bigger ones need 3–4in/7.5–10centimeters.

## *Nicotiana*

TOBACCO PLANT

***Solanaceae*** (Nightshade family).

***Scent:*** Rich and sweet, intensifies in the evening.

***Description:*** Nicotianas are a genus of about 30 species of annual or perennial shrubs or plants that bear tubular flowers of white, white tinged with violet, crimson or pale lime-green. The most familiar species is *Nicotiana tabacum* whose dried leaves are used for making commercial tobacco. It was named for Jean Nicot, Ambassador to Portugal in 1560, who promoted the powdered dried leaves as a snuff. Most gardens on the Iberian peninsula grew *N. tabacum* and the word 'cigar' comes from the Spanish 'cigarral', meaning 'little garden'. The reddish flowers are sweet-scented but the scent of smoking leaves in a pipe was acridly described by King James more than 300 years ago as: '. . . a custom . . . hateful to the nose, harmfull to the brain, dangerous to the lungs and in the black stinking fume thereof . . . resembling the horrible Stygian smoke of the pit that is bottomless . . .'

***Species:*** *Nicotiana alata affinis* (syn. *N. affinis,* 6ft/1.8m) is a pretty if somewhat leggy plant. It bears star-like white and green flowers that open in the evening, projecting a honeyed, sweet perfume to attract the moths that penetrate their long corolla tubes to obtain nectar and thus fertilize the plant at the same time. *N. alata grandiflora,* the Jasmine Tobacco (4–5ft/1.2–1.5m), is a very elegant variety, bearing clusters of pure white flowers that open like candelabra for as long as they are in shade; *N. suaveolens,* the Australian Tobacco plant (small white umbrella shaped flowers; 2ft/60cm) is an excellent potted plant; *N. sylvestris,* the Argentinian Woodland Tobacco (drooping white flowers; 6ft/1.8m) has a wonderful freesia-like scent that becomes stronger as the evening wears on.

***Cultivation:*** Nicotianas are best treated as half-hardy annuals, raised from seed planted in early spring and pulled up in the autumn to make way for the following spring's planting. However, in temperate climates, roots of a nicotiana will usually survive and seedlings will come up year after year. Incidentally, these seedlings are more likely to do well and flower profusely if left (if possible) where they come up. Nicotianas thrive in a deep rich soil containing plenty of leaf mold and compost. Plenty of moisture and partial shade are also required.

## *Nigella*

LOVE-IN-THE-MIST, LOVE-IN-A-TANGLE

*Only its blue eyes meet us day by day,*
*Till half we wish the mists would blow away:*
*Who knows true love*
*Be sure he also knows*
*Love-in-the-mist.*

Nora Chesson, nineteenth-century poet

***Ranunculaceae*** (Buttercup family).

***Scent:*** Spicy, nutmeg-like (seeds only).

***Description:*** This is one of the most delightful of cottage garden annuals, with pretty blue flowers peeping through tangled, feathery foliage. Although the flowers are scentless, the aromatic seeds have a nutmeg-like fragrance and have long been used to flavor breads and cakes. In France, they are still used to add zest to certain cheese products and an old country name for the plant translates as 'nutmeg-plant'. The seeds may also be placed in muslin 'swete bags' for scenting linen and were once much used as an inhalant, to clear head colds and restore one's sense of smell. John Gerard wrote:

> *The seed parched or dried at the fire, brought into powder and wrapped in a piece of fine lawn or sarcenet, cureth catarrhs, drieth the braine and restoreth the sense of smelling unto those which have lost it . . .*

Native to Egypt, Africa and ancient Persia, seeds of Love-in-the-mist were imported to Europe

*Nymphaea* sp.

during Elizabethan times, where the plants rapidly spread. Perhaps one of the more quaint reasons for the seeds' widespread culinary use was that women thought they would make their breasts larger!

***Species:*** *Nigella sativa* (pale blue flowers; 18in/45cm) bears the scented black seeds that may be harvested and powdered for culinary or domestic purposes; *N. damascena* (cornflower blue and, occasionally, soft pink flowers; 18in/45cm) does not have scented seeds.

***Cultivation:*** Love-in-the-mist seeds may be planted in early spring into a well-drained bed or pot. The flowers are pretty for indoor decoration and the seed, when harvested, should be placed in muslin bags and set among clothes.

## *Nymphaea*

WATER LILY

***Nymphaceae*** (Water Lily family).

***Scent:*** Exotic, heavy.

***Description:*** Water lilies hold pride of place in a water garden; they are the aristocrats of aquatic plants. They also have an interesting historical background, especially in ancient Egypt where they were extensively cultivated for temple ceremonies and for use in the home. On visiting the home of an Egyptian noble, guests would be presented with a single perfect lotus flower, which they were expected to either wear on their head or cup in their hands during the evening, indicating they came in peace. Water lilies were also prevalent at funerals of pharaohs; priests would lay wreaths of the flowers on the sarcophagus, from the chin downwards. Some of the wreaths are so well-preserved that botanists have been able to identify the individual species, which are largely the same as those that grow in Egypt today.

Water lilies were also sacred to the Buddhists. They regarded the flowers as a symbol of regeneration and purification as they rose, pure and straight, with the returning rains from the mud of dried-up watercourses. For many years water lilies had economic as well as spiritual importance. The starchy seeds were ground for food and the tubers were roasted and eaten rather like potatoes. Some species were used in fabric dyes, others as medicine for intestinal disorders. Nicholas Culpeper, writing of the white flowered *Nymphaea alba,* said: '. . . the leaves both inwards and outwards are good for agues, the syrup of the flowers produces rest and settles the brain of frantic persons. The distilled water of the flowers is effective for taking away freckles, spot, sunburn, and morphew from the face and other parts of the body. The oil of the flowers cools hot tumours, eases pains and helps sores.'

***Species:*** *Nymphaea caerula,* the Blue Lotus of the Nile (soft blue, narrow petaled flowers; hyacinth-like scent); *N. odorata* (white and pink flowers); *N. odorata* 'Sulphurea Grandiflora' (large yellow flowers) or *N. capensis,* the Africa Royal Purple Water Lily (rich plummy blue flowers). There are also many lovely hybrids—white, pink and yellow flowers in either large-flowered or miniature forms—largely due to the work of a certain M. Latour-Marliac. Known as the 'Father of the Water Lily', Latour-Marliac devoted his life to crossing and re-crossing many water lily species.

***Cultivation:*** All water lilies should be set in rich, firm compost in still water; avoid organic materials as they tend to turn the water green. Covering the compost with fresh shingle regularly will stop fish from rooting about in the soft mud and dirtying the water. Similarly, try to clip spent leaves and flowers before they sink in the water and rot. All water lilies love bright sunshine and require plenty of light in order to flower. The important thing to remember with these plants is that they must be planted in water of an adequate depth: *N. alba* needs 10ft/3m of warm water and other species need at least 2ft/60cm, preferably more. *N. caroliniana* 'Nivea' needs only 12in/30cm of water while *N. pygmea* 'Helvola' needs no more than 6in/15centimeters. Burying slow release fertilizer pellets in the compost material is also advisable.

## *Ocimum*

BASIL

*The smell of basil is good for the heart and for the head, that the sede cureth the infirmities of the heart, taketh away sorrowfulnesse which cometh of melancholy and maketh a man merrie and glad.*

John Gerard, *The Herball,* 1597

***Lamiaceae*** (Mint family).

***Scent:*** Pungent, slightly clove-like.

***Description:*** A group of aromatic annual herbs native to Asia and tropical Africa. Basil reached England in the middle of the sixteenth century and received a mixed reception. Most agreed on its medicinal and culinary uses but there remained a lingering suspicion, possibly because native Africans would chew the leaves to enable them to see visions before their various rituals. Basil was traditionally used as a strewing herb to keep away flies and Parkinson recommended it to be used '. . . to make sweet washing water . . . and sometimes [be] put in nosegays to procure a cheerfull and merrie heart'. It was also used in recipes for snuff.

In Hindu homes, basil is a sacred plant and protects against evil; every Hindu is cremated with a basil leaf placed on his or her chest. A sun-loving herb, basil is the perfect complement for summer salads and vegetables, especially tomatoes. The

*Ocimum basilicum*

drier the weather, the more leaves a basil plant will produce and these may be harvested for use in potpourri, aromatic vinegars or pasta sauce.

***Species:*** There are several distinct types of basil, including *Ocimum basilicum* (whorls of white flowers; 2ft/60cm) or the common Sweet Basil that has two types: mint or lemon-scented. Other varieties include the low growing Bush Basil *Ocimum b.* 'Minimum'; the attractive purple *O. b.* 'Dark Opal', and *O. basilicum* 'Citriodorum', which has soft pinky-green flowers and a powerful lemony scent.

***Cultivation:*** Culpeper recommended that basil '. . . be sown late . . . it being a very tender plant', and his advice holds true today. Set seed ¼in/6mm deep in planter trays indoors in a gentle heat, for basil will not germinate in cold ground. Germination will take 2–3 weeks. Another good idea is to plant basil in terracotta strawberry planter pots and keep them in a sunny place until the plants are ready to be thinned and replanted in the garden. Writing in 1613, herbalist Gervase Markham also thought this a good idea: 'Basil is sowen in gardens in earthen pots'. Basil is actually a better pot herb than a garden plant and will flourish on a warm window sill. Pinching out new flowers will result in a bushier plant. And, for the superstitious, folklore maintains that basil should never be planted near rue, as neither will prosper.

## *Odontoglossum*

***Orchidaceae*** (Orchid family).

***Scent:*** Rich, exotic; sweet or floral overtones, depending on species.

***Description:*** Odontoglossums are an extremely beautiful group of species and natural hybrids of orchids, which hail from Central and South American tropical rainforests. The flowers are large and richly fragrant, with a very wide range of colors, patterns and shapes; they are often 'crested'. Odontoglossums flower profusely—one stem can bear five blooms at once—and the flowers are long-lasting, making them a popular choice for either a greenhouse specimen or in cut flower arrangements.

***Species:*** *Odontoglossum pulchellum* (white and yellow flowers, sweet vanilla-like fragrance).

***Cultivation:*** General principles for growing odontoglossums are the same as for growing most orchids: a well-ventilated atmosphere, plenty of light and protection against overly harsh light. In addition, odontoglossums tend to prefer a somewhat cooler atmosphere than other orchids (60–65°F/16–18°C). Watering should be sufficient to keep the compost mixture just moist. Never overwater and ensure the potting mix is well combined, with drainage materials like coarse sand worked through well.

▶ A rich display of magnolia and dogwood.

### SCENTED FOOT BALM

Tired feet make for a tired-looking face. Pay your feet some attention and perk them up with a fabulous foot massage. Start at the toes and work up to your ankles and calves. Massage gently but firmly with a favorite rich moisturizer, such as this scented foot balm, paying special attention to the reflex zones under the arch of the foot and between the toes. After massage, rest and relax them for 10 minutes, lying with your feet raised above your head.

❖

*2 tbspns oil of lavender*
*2 tbspns oil of rosemary*
*2 tbspns oil of mint*
*1fl oz/25ml almond oil*
*3 tbspns lanolin*
*3 tbspns glycerine*

*Combine all the ingredients in a double saucepan, stirring continuously until evenly blended. Pour into a sterile, lidded jar and keep in the refrigerator.*

*Odontoglossum* sp.

## *Oenothera*

EVENING PRIMROSE

***Onagraceae*** (Evening Primrose family).

***Scent:*** Sweet, with slight lemony tang; strongest in the evening.

***Description:*** The Evening Primrose was originally native to North America and the West Indies. The most familiar species, *Oenothera biennis,* was imported to Italy in 1619 and from there to England with John Parkinson recording cultivation details for a 'tree primrose' in 1629. A biennial plant with a delicious perfume, the evening primrose grows to a height of 3-4ft/1-1.2m and bears large (to 3in/7.5cm across) yellow, cup-like flowers continuously throughout summer to autumn. These flowers are pollinated by moths, so they only open at night, hence the name 'evening' primrose. During the day, the flower petals are held in place by protective sepals that hook together, rather like a hood. As soon as the cool of evening descends, the flowers open quickly, prompting John Keats to comment that he was '. . . startled by the leap of buds into ripe flower . . .'

The botanical name is derived from the Greek *oinos* (wine) and *thera* (a hunt) and refers to its possible use in dispelling the ill effects of wine. The roots of the plant may be eaten and the French often use the flowers as a garnish. A pretty idea is to float them in fingerbowls for a summertime twilit supper.

***Species:*** *Oenothera odorata* (yellow flowers 3-4in/7.5-10cm across—2ft/60cm in height) is the most powerfully scented variety. Also lovely in a scented garden, either at the front of a garden or tucked in shady corners to contrast with darker foliage, are *O. biennis* (yellow flowers; 4–5ft/1.2–1.5m); the dwarf *O. caespitosa* (white flowers; 12in/30cm); *O. californica* (white and yellow flowers; 2ft/60cm) and the prostrate form, *O. missouriensis* (yellow flowers).

***Cultivation:*** Being reasonably hardy, evening primroses will grow in most soil types and conditions. Sow seed in midsummer and, from then on, they should self-sow. After flowering, harvest the seeds and scatter them in odd spots throughout the garden, especially under windows, so the scent may be readily enjoyed next year.

## *Origanum*

MARJORAM

> *It is a low shrubbie plant, of a whitish colour and of marvellous sweet smell . . . the stalks are slender . . . about which grow forth little leaves soft and hoarie. The flowers grow at the top (and are) of a white colour. The whole plant and everie part thereof, is of a most pleasant taste and aromatic smell . . .*
>
> John Gerard, *The Herball* (1597)

***Lamiaceae*** (Mint family).

***Scent:*** Sweet and slightly spicy.

***Description:*** The marjorams are a group of annual herbs native to the Mediterranean region. The botanical name *origanum* means 'joy of the mountains', for it grew wild in the Greek hills. Marjoram was used extensively by the ancient Greeks as a medicine as well as a condiment. They also believed that if marjoram grew on a grave then the dead person was happy. An old funeral prayer exhorted: '. . . many flowers [to] grow on this tomb, violets and marjoram and the narcissus growing in water and around thee may all Roses grow.'

Medieval herbalists advised that smelling wild marjoram frequently would keep a person in good health. It was much in demand as a strewing herb and an anonymous seventeenth century courtier tells us that his lady's pew was '. . . strewn full gay with primroses, cowslips and violets, sweet with mints, with marigolds and with marjorams . . .'

Marjoram has always been included in 'swete bags' to be laid among linen and in potpourri, and the leaves were rubbed over wooden furniture to polish it and impart a spicy perfume. Being sweet and slightly sharp, marjoram complements rose and lavender especially well.

Marjoram has many culinary uses. Izak Walton suggested it be sprinkled over freshly caught pike, along with thyme and winter savory. It goes very well with salad vegetables, especially tomatoes, and is a key ingredient in *bouquets garni.* It was once much used to make jam and a pretty recipe to experiment with comes from *The Receipt Book of John Nott,* cook to the Duke of Bolton in 1723:

A CONSERVE OF MARJORAM

> *Take the tops and tenderest parts of Sweet Marjoram, bruise it well in a wooden Mortar or Bowl; take double the weight of Fine Sugar, boil it with Marjoram Water till it is as thick as Syrup, then put in your beaten Marjoram.*

Marjoram tea, made by infusing a handful of the leaves in a pint of boiling water, was once a popular drink among country folk. So was marjoram 'milk' (milk in which a muslin bag of marjoram had been suspended for a time) and the herb was also used to flavor beer.

***Species:*** *O. marjorana,* or Sweet or Knotted Marjoram (pale mauve flowers; 9in/25cm); *O. onites* or the Pot Marjoram, so named because it was grown indoors in pots so cooks could readily access its strongly flavored foliage (white flowers; 12in/30cm); and the Wild Marjoram, *O. vulgare* (pinkish mauve flowers; 2ft/60cm), which Lyte's *Herball* lists as a 'noble and odoriferous plant'.

***Cultivation:*** Marjorams are somewhat tender and require plenty of sun and a light, dry soil. Only the Sweet Marjoram (*O. marjorana*) seems able to survive a really cold winter; the rest should be wintered indoors on a kitchen window sill. Marjorams may be readily grown from seed planted in spring and stock may be increased by root division or taking cuttings through autumn. Marjorams are well loved by bees and butterflies and will attract them to your garden.

SCENTED PINCUSHION

Make sewing even more enjoyable by slipping a fragrant sachet into your pincushion. Every time you press a pin or needle home, a puff of scent will be released, perfuming both the work and the sewer. A pretty mix for such a sachet is:

*1oz/25g rose petals*
*1oz/25g lemon verbena*
*1 tspn marjoram leaves*
*1 tspn cloves, crushed*
*1 tbspn orris root powder*

*Lavender may be substituted for the lemon verbena, if desired.*

## Olearia

MAORI HOLLY, DAISY BUSH

***Asteraceae*** (Sunflower family).

***Scent:*** Musky.

***Description:*** The olearias are pretty evergreen trees and shrubs that are native to Australia and New Zealand. They are commonly known as Daisy Bushes, being covered with clusters of fragrant flowers that are usually white. They are extremely hardy and do well in coastal areas. Leaves from the New Zealand Daisy Bush, *Olearia angustifolia,* were worn by Maori tribespeople as a sign of respect for the relatives of the newly dead.

***Species:*** *Olearia macrodonta* (large white flowers; 20ft/6m) has toothed, holly-like foliage, as does the smaller Maori Holly, *O. ilicifolia* (yellow and white daisy-like flowers; 10ft/3m). The hardiest is *O.* x. *haastii* (clusters of tiny white flowers; 6ft/1.8m). All are tolerant of salty air and dry, poor soil; all emit a powerful musky fragrance from both flowers and leaves.

***Cultivation:*** Olearias require a well-drained soil and a sunny, sheltered position. They are ideal for planting in a seaside garden, a shrub border or as a windbreak. No regular pruning is needed other than trimming dead wood; they may be easily propagated from cuttings taken in summer.

## Paeonia

PEONY

*Double and single paeonies are fit flowers to furnish a garden and by reason of their durability give one fresh pleasure every year without any further trouble.*

John Parkinson

***Paeoniaceae*** (Peony family).

***Scent:*** Distinctly rose-scented; some species, however, have an unpleasant odor.

***Description:*** This perennial cottage garden plant is steeped in history. It was thought to have been named after the Greek physician Paeos, who used the root extract to treat Pluto's wounds. Every part of this plant has some medicinal virtue: the seeds are used as a restorative tea, the flowers for a wine or syrup and the kernels may be eaten as a spicy condiment and were once used as a decorative garnish in much the same way as sliced almonds. The ancients held the plant in some awe and believed it to be an emanation of the moon because the seeds of certain species have a phosphorescent quality and they glow at night-time. Much ritual accompanied harvesting of the peony's roots; they had to be dug at dead of night

*Paeonia moutan*

for if the operation was viewed by a woodpecker, the digger risked possible attack and loss of his eyesight!

The Chinese called the peony *shaoyao*, which means 'charming', and the flowers were exchanged as tokens of love and friendship. The most popular peonies, *Paeonia lactiflora* (syn. *P. albiflora*), are often known as Chinese Peonies because the first hybrids were raised in that country. In fact, a Chinese monograph written in 1066 records that different people would travel from 'the hills around Loyang' to sell a variety of cuttings to the townspeople for grafting. During the T'ang Dynasty, emperors placed peonies under their protection and they were even offered as marriage dowries because they were so valuable. In ceramic, textile and pictorial art, peonies were usually depicted with a phoenix, it being deemed fitting that the King of Flowers should keep company with the King of the Birds. Peonies were introduced to England and Europe from China and, in Tudor times, contemporary writers mentioned them frequently, often using quaint country names, such as 'Chesses', 'Hundred Bladed Rose', 'Sheep Shearing Rose', 'Pie Nanny' and 'Marmaratin'. John Parkinson, in his *Paradisus,* wrote '. . . we cherish them for their beauty and delight of their goodly flowers as well as for their physical virtues.'

## THE COTTAGER'S POTPOURRI

A charming blend, pleasing to the eye and with a long-lasting fragrance. Display it in an antique lidded sugar bowl, pretty china serving dish or silver tray.

❖

*3½oz/100g peony petals*
*2oz/50g pink rose petals*
*1oz/25g chamomile flowers*
*1oz/25g pink carnation petals*
*1oz/25g borage flowers*
*1oz/25g lemon verbena*
*1oz/25g hyssop*
*1oz/25g lemon thyme*
*½oz/15g mint, or mixed mints*
*½oz/15g southernwood*
*½oz/15g sweet woodruff*
*1oz/25g dried marigolds, or everlasting daisies, for color*
*1 tbspn allspice powder*
*1 tbspn cinnamon powder*
*1 tbspn cloves, crushed*
*1 tspn nutmeg, grated*
*1 tbspn dried lemon zest*
*2–3 tbspns orris root powder*
*essential oils—rose, carnation*

❖ ❖ ❖

***Species:*** *Paeonia lactiflora*—white flowers with crimson shoots and many-budded stems; pronounced rose perfume) has dozens of hybrids, most of which offer incredible perfume and a variety of flower shapes and sizes. *P. mlokosewitschii,* the Lemon Peony, is a beautiful Caucasian species (large single yellow flowers; sweetly perfumed; 2ft/60cm). Hybrids of the lovely shrub *P. suffruticosa* (*P. moutan*) also bear very large flowers, all of which have scent. The latter are often called Tree Peonies and most have a lovely yeasty sweet perfume. Tree Peonies should always be planted in a sheltered position as they have slender stems that snap easily in wind. In the Orient, Tree Peonies have a long history; the Japanese are expert at grafting them and, in other parts of Asia, the flowers are picked and eaten as a vegetable.

***Cultivation:*** Peonies are lime-tolerant and greedy feeders, requiring plenty of well-rotted compost or manure and a handful of bonemeal both before and after flowering. They are hardy and will tolerate frost in winter, but require a position sheltered from the wind. The chief cause of failure is planting peonies too deep or in tightly packed soil; the root crown should be no more than 2 inches below the ground level and take care to spread the roots out well before covering with earth. Peonies live a long time in gardens when grown under suitable conditions. Some records state single plants existing for more than a hundred years. Most species also thrive as potted specimens and can be forced, under greenhouse conditions, into flowering from spring through to autumn.

*Passiflora caerulea*

## *Passiflora*

PASSIONFLOWER

***Passifloraceae***

***Scent:*** Delicate, honeyed; overtones of orange.

***Description:*** Passionflowers are a group of herbaceous plants and shrubs, mostly climbers, which have very attractive foliage and wonderfully intricate flowers that often have a sweet scent as well as egg- or cucumber-shaped fruit of excellent flavor. They were named 'Passionflower' by the sixteenth century Jesuit priests who traveled to South America. The story goes that when they stepped ashore and saw a *Passiflora caerula* vine, they took the plant as an omen that Catholicism would be the religion of this 'New World', the ten sepals represented the ten apostles, the inner corona the Crown of Thorns, the five stamens being Christ's five wounds, the three-parted stigma the nails, and the ovary the hammer with which the wounds were made.

***Species:*** *Passiflora caerulea,* native to Brazil (white flowers with pink-tinged petals and purple centers up to 4in/10cm across; 12–15ft/3.6–4.5m) is the best known species, being a vigorous climber with what must surely be among the most interesting and beautiful flowers in the plant world. The giant *P. quadrangularis* (4in/10cm flowers of alternating white sepals and magenta petals; richly fragrant) bears long thick fruit filled with sweet purple juice. It is a magnificent sight in a wild tropical garden, weaving through the trees, and is also a satisfactory greenhouse specimen. Cultivated for its edible fruits is *P. edulis* (white or purple-tinted flowers). A spectacular climber for warm outdoor conditions, such as Australia, is *P. mollissima,* the Banana Passionfruit. This species bears masses of delicately scented rose-red flowers, which give way to yellowish-white cucumber-shaped fruit.

***Cultivation:*** Passionflowers are slightly tender. They will grow exuberantly in warm climates but need to be kept in heated greenhouses in less favored areas. They require a well-drained soil with plenty of water during spring and summer, but are not overly fussy about the soil quality. More important is the aspect; an ideal choice is to train vines on supports against a warm sunny wall, or over a pergola. Passionflowers are a great favorite with bees and also with birds; netting might be a good idea if you want to preserve your fruit harvest.

## *Paulownia*

CHINESE FOXGLOVE TREE

***Scrophulariaceae*** (Figwort family).

***Scent:*** Warm, incense-like.

***Description:*** These are small but extremely striking ornamental trees, native to China and Japan, with large clusters of rich purple or blue foxglove-like flowers that emit a luscious exotic fragrance in early spring. They were revered in ancient Japan (much like the chrysanthemum) and featured on the imperial crests of the Mikado. The wood was once an important commercial item, being used to make musical instruments and furniture; it was also charred and ground to produce gunpowder.

***Species:*** *Paulownia fargesii* (purple and yellow flowers; 20ft/6m); *P. tomentosa* (*P. imperialis*—violet flowers with blue tinges; 40ft/12m).

***Cultivation:*** Paulownias are exquisite ornamental trees for a small garden. Even when they are not in bloom the plush-like foliage makes a handsome foil for other plants. They should be planted in a position of full sun, which is reasonably well sheltered. They are not frost tolerant and can often be caught out by their early flowering habit in a cold climate.

## *Pelargonium*

*On the backs of my armchairs are thin Liberty silk oblong bags, like miniature saddle bags, filled with dried lavender, sweet verbena and sweet geranium leaves. This mixture is much more fragrant than the lavender lone. The visitor who leans back in his chair wonders from where the scent comes.*

Mrs Earle, *Potpourri from a Surrey Garden*, 1905

***Geraniaceae*** (Geranium family).

***Scent:*** A varied range.

***Description:*** The Pelargonium group contains a large number of shrubby perennial plants bearing both single and double flowers in a wide range of colors including pink, white, purple, and scarlet. The group includes the zonal, or well known bedding geraniums, along with variegated-leaf or ivy-leafed varieties. However, the most important varieties to concentrate on here are the Scented Leafed Pelargoniums. These plants were discovered in the Cape Province of South Africa and brought to England during the time of Charles I. Their popularity soared during Victorian times when they were cultivated in fashionable glasshouses built by the newly rich. They were also popular as potted house plants and would often be placed up the side of stairs so the women's long skirts would brush against them as they passed, thus releasing their perfume.

Scented pelargoniums soon became a familiar feature in cottage windows. In these far humbler surroundings, many uses were found for the leaves: in relaxing hot baths, for pillows, sachets and potpourri and for adding scent and flavor to custards, jams and jellies. Miss Gertrude Jekyll whimsically referred to the minty leaves of *Pelargonium tomentosum* as being 'as thick as a fairy's

*Pelargonium* x *domesticum* hybrid.

blanket' and most suitable for making a delicious jelly. The rose-scented leaves of *P. capitatum* give a light pleasant flavor to creamy desserts and puddings and are a pretty garnish for jams and jellies. An oil extracted from this species, along with several others, is used as an adulterant for costly attar of roses and is the source of perfume for most common 'rose'-scented cosmetics.

***Species:*** Specialist nurseries list all kinds of varieties, and perfumes range from lemony and rose, to eucalypt, mint, apple—even ginger and coconut. Some of the most popular are rose-scented pelargoniums (*Pelargonium capitatum* and *P. grossularioides*); fruit-scented pelargoniums (*P. odoratissimum* (apples) and *P. citriodorum* (oranges); *P. nervosum* (lime) and the peppermint scented *P. tomentosum*). The lemon-scented *P. crispum* was much used as a decoration for finger bowls and, because the leaves retain their perfume for some time, is also most suitable for potpourri. More unusual scents come from *P.* 'Clorinda' (eucalyptus), the pungent *P. quercifolium* or Oak-Leafed Geranium (incense-like) and *P. denticulatum* (balsamic).

***Cultivation:*** Scented Leafed Pelargoniums are quite easy to propagate from cuttings and should be planted out in full sun and in a well drained, compost-rich soil. Fairly dry conditions seem to produce more flowers, with excess moisture resulting in more leafy shoots. Ideally, they should be planted by the edges of a sunny path so they may be brushed against by passers-by. They are vulnerable to frost so, in cold climates, they are best wintered indoors as potted house plants.

## *Phlox*

***Polemoniaceae*** (Phlox family).

***Scent:*** Sweet and musty; a little like freshly mown hay.

***Description:*** Phlox is a group of hardy perennial flowering plants, native to the eastern states of America, which add splendid color and fragrance to borders and rockeries. They are also lovely sown in drifts around shady trees. Phlox were introduced to England and Europe by John

*Phlox subulata*

Bartram in 1745, as '. . . one sod of the fine, creeping Lychnis' and rapidly became established as favorite cottage garden plants. This 'Lychnis' was the Moss Phlox or *Phlox subulata,* one plant of which can cover several square feet with a carpet of pink, mauve or white flowers. However, it is the many showy hybrids and cultivars, with their striking colors and hardy, accommodating habit, which we are most familiar with today.

***Species:*** *Phlox paniculata* or Autumn-flowering Phlox (lilac, purple or white flowers; 4ft/1.2m); *P. subulata,* the Spring-flowering Phlox, or Moss Phlox (usually soft pink flowers; creeping habit); *P. bifida* (pale mauve flowers); *P. maculata* (mauve/purple flowers on tall stems; 2ft/60cm); the creeping Evergreen Phlox, *P. stolonifera,* and the American Wild Sweet William, *P. divaricata,* which has showy blue flowers. There are many other varieties, all with vibrant colors and varying amounts of scent such as *P.* 'Temiskaming' (plum-colored flowers); *P.* 'Brigadier' (flame red flowers) and the lovely pure white *P.* 'Frau Buchner', which has been a popular bedding plant since its introduction in 1910. The dainty *P. drummondii* 'Twinkling Stars', with its star-shaped corollas, makes a pleasing contrast to the round phlox varieties.

***Cultivation:*** Seedlings should be planted in late autumn in a rich soil. They prefer medium shade and cool conditions and require plenty of water during their growing period. Phlox should be thinned out to ensure better flowering and stronger individual plants. Ideally, every few years the roots should be lifted and replanted into new soil; this is also the best time to increase stock by taking cuttings. Root cuttings will also help avoid eelworm, a pest that phlox seem to be particularly prone to.

## *Pieris*

ANDROMEDA TREE OR LILY OF THE VALLEY BUSH

***Ericaceae*** (Heather family).

***Scent:*** Penetrating; like mignonette.

***Description:*** This group of very attractive evergreen ornamental shrubs, with their drooping sprays of papery bell-like flowers in spring, are native to Japan and the Allegheny Mountains of the United States. As a bonus, when flowering is ended, the oblong tapering leaves sprout shining amber-colored shoots, giving the appearance of a second 'flowering'.

***Species:*** *Pieris floribunda* (pure white, bell-shaped flowers' upright sprays of 4-6ft/1.2-1.8m), or Andromeda, was originally named for the beautiful daughter of the Ethiopian King, Cepheus; *P. japonica* (white flowers; drooping sprays of 10ft/3m) and the beautiful Himalayan *P. formosa* (white flowers and rosy-red spring leaves; 20ft/6m sprays).

***Cultivation:*** Pieris may be planted in cool moist places and requires a fairly acidic soil, preferably with sand and/or peat moss mixed through. Plenty of shade is best, with sun for flowering.

*Pieris japonica*

*Plumeria rubra*

## *Plumeria*

FRANGIPANI

***Apocynaceae*** (Dogbane family).

***Scent:*** Sweet and rich, more intense in the evening.

***Description:*** These ornamental trees and shrubs, native to Mexico and the West Indies, have become the best loved plants of tropical lands and most temperate areas. Legend has it that the plant was named for an Italian Count Frangipani, who distilled the flowers' volatile oils and combined them with other essences before presenting his 'Frangipani perfume' to the noble ladies of Europe. Catherine de Medici was said to have worn no other fragrance and set the fashion for fine white kid gloves imbued with the scent; the Count's fame and fortune was assured from then on.

Bearing immense clusters of waxy flowers with the most delicious scent, the plants were frequently placed by Buddhist shrines and temples in Asia where it was known as the Pagoda Tree or Temple Tree. Frangipanis are a funereal flower in Malaysia, where Muslims place them on graves of the newly dead. The plant's ability to flower, even when lifted out of the soil temporarily, explains its association with immortality and the afterlife.

The flowers are dried in some Asian countries and put among clothes in much the same way as lavender. In Java, the jellied flowers are a popular sweetmeat. Although the sap is poisonous, it was once used medicinally by Amazonian natives. In his *A Naturalist on the Amazon* (1848), explorer Henry Bates recorded that:

> *. . . one of the most singular of the vegetable production of the campas is the Sucu-u-ba tree (frangipani) . . . the bark and leaf stalks yield a*

*copious quantity of milky sap which the natives use very generously as a plaister in local inflammation, laying the liquid on the skin with a brush and covering the place with cotton. I have known it to work in many cases . . .*

***Species:*** *Plumeria rubra* f. *acutifolia* or Pagoda Tree is the old-fashioned yellow-and-white favorite (clusters of creamy flowers 9-10in/20-25cm across; 13-20ft/4-6m), which is so common in older working class suburbs of Australia, probably because it could be struck so easily. The Burgundy Flowered Frangipani, *P. rubra* (red/pink flowers; 18ft/5m), is native to Martinique, where the women call it Red Jasmine and wear the rosy flowers in their hair.

***Cultivation:*** Frangipanis are deciduous and adapt to most soils and conditions. In tropical areas, a tree will flower almost continuously throughout the year in an enriched soil. However, they are frost-tender and need hot-house conditions when in a cold climate. Frangipanis are delightful planted outside the house, preferably in front of a window that receives sun; not only do the smooth tapering leaves provide shade and help keep the house protected, but the perfume will waft inside to cool and freshen the air. Cuttings should be taken when the plant is not flowering. In temperate areas, the cutting should be prepared with commercial cutting powder and allowed to root before being planted out. In the tropics, however, blasé gardeners have been known to just snap off a stem and stick it in the ground, confident that nature will be capable of carrying on from there!

## Primula

PRIMULAS, PRIMROSES, AURICULAS, COWSLIPS

A FLOWER PUDDING

*Mince cowslip flowers, clove gillyflowers, rose petals and spinach of each a handful, take a slice of Manchet [white bread] and scald it with cream. Add a pound of blanch'd Almonds pounded small with Rose-water, a quarter of a Pound of Dates sliced and cut small, the yolks of three eggs, a handful of Currants and sweeten with Sugar. When boiled pour Rose-water over and scrape Sugar on. Then serve up.*

From *The Receipt Book of John Nott,* Cook to the Duke of Bolton, 1723

***Primulaceae*** (Primrose family).

***Scent:*** Sweet, mossy; varying in intensity and type, depending upon species.

***Description:*** The genus *Primula* is a large family, encompassing some 500 species, all native to the Northern Hemisphere and best suited to cool moist conditions in semi shade. Not all the plants are fragrant, but most of the old-fashioned strains of primroses, cowslips and auriculas are sweetly scented and add pretty patches of color to the garden. The European Woodland Primrose, *Primula vulgaris,* is emblematic of the English countryside and each spring the pleasantly scented clumps of yellow flowers bloom in gardens and hedgerows alike. Since earliest times, the flower has been a poet's favorite. Shakespeare used it both as a symbol of fragility, referring to the '. . . pale primroses, that die unmarried, ere they can behold Bright Phoebus in his strength', and as a lover's flower; witness Hermia in *A Midsummer Night's Dream*:

*And in the wood, which often you and I*
*upon faint primrose beds were wont to lie,*
*Emptying our bosoms of their counsel sweet,*
*There my Lysander and myself shall meet.*

The primrose's name is derived from *prima vera,* the Latin expression for the first flower of spring, and many references are made to its association with new beginnings and youthfulness. To Robert Herrick, the primrose was the 'Sweet Infanta of the Year' and in *Two Noble Kinsmen,* the primrose

▶ The mountain hue is tinged with the fragrance of scented plants.

*Primula japonica*

becomes the '. . . first born child of Ver, merry springtime's harbinger with her bells dim.'

More practical uses for primroses saw their inclusion in 'sallets' and 'potages'. In his *Acetaria* (1699) John Evelyn gave a recipe 'To Make a Grand Sallet for the Spring', which entailed taking 'all manner of Cowslips, Violets, Strawberries, Primroses, Watercress . . . [and] . . . Alexander-buds . . . either separately or apart' and arranging them in layers around a centerpiece like steps 'going up to a cross' before dousing the whole lot with a '(much approv'd) Receipt' for vinegar. Primrose wine and primrose pudding are still made in some English country districts and they remain a favorite flower for picking among adults and children alike, being particularly useful for a dainty arrangement, such as an old-fashioned posy of flowers. In addition to the classical single-stemmed yellow primroses, many different colored hybrids are available; *P.* 'Garryarde Guinevere' has crimson-colored, richly scented blooms.

Auriculas (*Primula auricula*) were alpine flowers, with fragrant yellow or white blooms on stout stems. They have been cultivated since the sixteenth century, when John Parkinson noted that '. . . auriculas do seem every one of them to be a nosegay alone of itself . . . their pretty scent doth an increase of pleasure in those that make them ornaments for their wearing.' They are very popular with hybridists who have produced many new types in different colors: florists' auriculas come in red, crimson, purple and salmon, but the perfume tends to be most evident in the yellow or white originals.

The exquisite little cowslip (*Primula veris*) is found growing wild in water-meadows in Europe and can be readily grown from seed in a moist, sunny spot in the garden. Cowslips emit a soft fragrance, which always arouses admiration; Shakespeare described the scent as a blend of burnet and clover while Sir Francis Bacon recommended cowslips be planted alongside paths with chamomile and thyme where '[they] will perfume the aire most delightfully'. Indeed, the old country name 'cowslip' is a reference to the sweet, gentle scent, with 'cuslippe' being the old English term for the 'breath of a cow'. Cowslips once had many culinary uses: the leaves were minced and tossed with lettuce in 'sallets', the flower buds were infused for tea and a lovely wine, made from the petals, could be taken for insomnia. A quaint game for young lads and lasses of yore was tossing a 'ball' made of tightly pressed cowslip flowers from one to another with the accompanying rhyme:

*Tisty, tosty, tell me true*
*Who shall I be married to?*

The person who caught the ball as it fell apart would be the enquirer's future partner.

***Species:*** *Primula vulgaris* (syn. *P. acaulis*), the Woodland Primrose (pale yellow flowers; 3in/7.5cm); *P. japonica*, the Japanese Primrose (white, pink or mauve flowers; 1ft/30cm); *P. pulverulenta*, the Chinese Silverdust Primrose (deep red flowers; 1ft/30cm). Also look for some of the old-fashioned double primroses like 'Red Paddy' and 'Crathes Crimson', the stunning 'Tyrian Purple' and the rose-pink 'Rose du Barri'. Deliciously perfumed auriculas are divided into two main groups: the edged varieties and the self colors. Within this enormous selection, 'Blue Velvet' (purple or blue flowers with honey scent) and the 'Dusty Miller' series (red, purple and yellow flowers) are the most fragrant. The main cowslip you will find in nurseries is *Primula veris* (yellow flowers; 18in/45cm). Also worth pursuing through a specialist seed supplier is the Giant Himalayan Cowslip (bright yellow fragrant flowers; 3ft/1m).

### PRIMROSE PUDDING

❖

*10–12 primroses*
*10½fl oz/300ml milk*
*2oz/50g caster (superfine) sugar*
*4 tbspns rice meal*
*4 tbspns ground hazelnuts*
*10½fl oz/300ml double cream*

*Infuse the flowers in the milk for 15 minutes; strain. Combine sugar, rice meal and nuts. Add milk slowly over a gentle flame, stirring until the mixture thickens. Add half the cream, pour into dessert bowls and refrigerate. Whip the remaining cream and pipe it around the bowls. Decorate with crystalized primrose flowers.*

***Cultivation:*** All members of the primula family, being woodland plants, require cool conditions and plenty of moisture. Cowslips should have a more open position than the other species. They should all have a rich, loamy soil and, as they tend to form new roots on the soil surface, should be regularly mulched. Mushroom compost is the best. Primroses, cowslips and auriculas may all be propagated readily from seed. All are valuable as bedding plants but it is the self-set seedlings that bring charm to a garden, especially if they fall in drifts about trees and shrubs.

## *Prunus*

ORNAMENTAL FLOWERING FRUIT TREES

***Rosaceae*** (Rose family).

***Scent:*** Delicate, honeyed; overtones of almond or peach, depending on species.

***Description:*** A group of evergreen or deciduous trees and shrubs with aromatic leaves and scented flowers, the prunus family includes the Flowering Almond, Apricot, Cherry, Peach and Nectarine. A number are grown for garden decoration, not just for the exquisite pale spring blossom, but for the attractively colored foliage later in the year. Not all species are perfumed.

***Species:*** *Prunus* x. *yedoensis*, the Yoshino Cherry (almond-scented white flowers; 20–25ft/6–8m) is a wide-spreading tree, ideally suited for placement in the center of a lawn and underplanted with spring-flowering bulbs. *P. laurocerasus*, the Cherry Laurel (honey-scented creamy flowers; 20ft/6m) was once very popular in Victorian shrubberies and is frequently seen in Europe as a hedging plant. The most popular varieties of Japanese Flowering Cherries include: *P. serrulata* 'Amanogawa' (semi double, soft pink flowers; 20ft/6m); *P. s.* 'Ukon' (yellowish green flowers; 20ft/6m); *P. s.* 'Shirofugen' and *P. s.* 'Shirotae'

*Prunus* sp.

(both double white flowers). *P. lusitanica,* the Portuguese Laurel (drooping white flowers; 20ft/6m) has long evergreen myrtle-like leaves and is an attractive year-round garden specimen for its blossoms are followed by decorative red/purple oval fruits. Flowering apricots have been widely planted for many centuries. The Southern European Apricot, *P. armeniaca,* was plentiful in the Holy Land and its fruits were described in the Bible as being like 'Apples of gold in pictures of silver'. Specimens of the Chinese Flowering Apricot (*P. mume*), which are thought to be more than a thousand years old, may be seen in Asian temple gardens. The blossoms have a very sweet scent although, traditionally, these trees have been primarily cultivated for their fruit.

***Cultivation:*** Prunus trees prefer a well-drained, slightly alkaline soil in full sun or light shade. Most are hardy and, although initially slender when young, will spread as they grow, so do not plant where space is limited. I have one growing in the border that runs along the driveway and its enthusiastic growth prevents me from opening the car door, for it has spread to 6ft/2m wide! Judicious pruning in the spring, when the sap is rising, will help to shape these plants.

## *Reseda*

MIGNONETTE

***Resedaceae*** (Reseda family).

***Scent:*** Sweet and powerful.

***Description:*** This is a small group of annual plants, mainly native to the Mediterranean regions and to North Africa. The botanical name, *reseda,* comes from the Latin phrase meaning 'to assuage', for it was thought the plants could soothe a troubled mind and heart. Only one species is fragrant, the sweet-scented Mignonette, which was

*Reseda odorata*

first introduced to Europe in the early part of the eighteenth century by Napoleon, who sent the seed to his love, Josephine. She grew the inconspicuous little brownish-yellow flowers in small pots, placing them around her apartments to release their perfume. It became an extremely fashionable plant, being grown in large quantities to provide cut blooms for winter posies. The French named it mignonette, meaning 'Little Darling', hence Cowper's reference to: 'The fragrant weed, the Frenchman's Darling.' Mignonette was often grown in large tubs or pots, which were placed on the balconies outside elegant apartments in London or Paris, thus masking the stench rising from the streets below. In 1830, the historian Phillips noted that city dwellers '. . . must be delighted with the scent it removes from the balconies into the streets, giving a breath of garden air to "close-pent man" whose evocations will not permit a ramble beyond the squares of the fashionable part of town'.

A floral vinegar made from fresh mignonette flowers makes 'a most excellent and refreshing liquor to smell at by those afflicted with headache', wrote Mrs Leyel. Mignonette vinegar will soften facial and body skin and give off a refreshing scent in a sick room. To clear skin irritations and even out patchiness, add 1 cup of mignonette vinegar to a warm bath. Also, after exposure to hot sunshine, dab a little behind ears and onto temples to refresh the spirits.

MIGNONETTE VINEGAR

❖

*1½ cups fresh mignonette flowers*
*2 cups white wine vinegar or apple cider vinegar*

*Put mignonette petals in a large, clean glass bottle. Gently warm the vinegar in a non-aluminum saucepan, then pour it over the petals and cap securely. Leave on a sunny window sill or by a warm stove for 2 weeks, then strain before using.*

***Species:*** *Reseda odorata* (compact yellowish-brown flower spikes; 2ft/60cm) is the most fragrant form and the flowers are distilled to extract an essential oil. Some more modern varieties offer orange, gold, crimson or cream blooms, but they are not as fragrant as the original species.

***Cultivation:*** Mignonette is hardy and may be readily propagated from seed in well-prepared potting mix with plenty of rich compost. Seed may be sown either in autumn for pots or in spring, if outdoors. Mignonette makes a charming edging plant for the summer months, as well as a potted specimen on a sunny window sill.

## *Romneya*

CALIFORNIAN TREE POPPY

***Papaveraceae*** (Poppy family).

***Scent:*** Exotic, sweet; strongest at evening.

***Description:*** Californian Tree Poppies are entirely different from other poppy species. Native to California, they are not easy to grow but, once established, form very attractive dense bushes each year. Californian Tree Poppies are perennial and grow quite tall (4ft/1.2m). They have blue-gray somewhat leathery leaves and very beautiful large flowers with white crinkly petals and a cheerful pincushion-like cluster of bright yellow stamens, which emit an exotic perfume. More fragrant at night, Californian Tree Poppies are a beautiful sight in a twilit garden when the white flowers seem luminous, creating an inviting glow in a shrubbery or dim corner of a patio.

***Species:*** *Romneya coulteri,* Californian Tree Poppy (large white fragrant flowers; 4ft/1.2m).

***Cultivation:*** Californian Tree Poppies do best in full sun and loose, gravelly soil. Abundant watering in summer is essential. They dislike being disturbed and are difficult to transplant, being best propagated from seed or root cuttings. Californian Tree Poppies, once they feel at home, will rapidly form a hedge or cover an unsightly wall. Consider planting a row as a scented 'screen' around a terrace or paved courtyard area. They are reasonably hardy and will form an exquisite fragrant backdrop for twilight entertaining.

## *Rosa*

ROSE

*Dry roses put to the nose to smell do comfort the Brayne and the herte and quickeneth the spyryte.*

Richard Banckes, *A little herball,* 1525

***Rosaceae*** (Rose family).

***Scent:*** Sweet; varying in intensity, dependent upon species.

***Description:*** This is a very large and important family, containing about 20 species and many thousands of cultivars. The history of the rose goes back many centuries; fossil remains of roses, estimated to be at least 35 million years old, have been found in Colorado. Our ancestors prized roses for their sweet scent and they have long been the most popular garden plant of all. Roses first came to Europe from Persia and the ancient Greeks and Romans set great store by the flowers. Not only were the floors of their palaces strewn with roses and guests crowned with them, but food and wine were flavored with the petals. Pliny wrote that meat was usually served with rose petals pressed all over it. At one lavish banquet, hosted by the Emperor Elagabalus (A.D. 218–222), so many rose petals were showered upon the dinner guests that several suffocated.

*Rosa* 'Iceberg'

Later still, roses came to have many symbolic meanings, the most popular being 'love', for rose perfume was said to be one of the favorite scents of Aphrodite and Venus. Roses were hung from the ceiling of a room where secret matters were discussed, giving rise to the term *sub rosa,* meaning 'privacy is assured'. The plaster 'roses' used as decorative motifs on ceilings in private homes are a remnant of this custom. When the days of the debauched Romans were over, the early medieval church adopted the rose as its own, claiming the white rose in particular, *Rosa alba,* as representing the Virgin Mary. Exquisite stained glass 'rose' windows and carved wood ornaments and candlesticks may still be seen in many old chapels. Even 'rosaries' were so named because the beads were once made from tightly compressed rose petals, which gave off their pretty scent when they were handled.

### DAMASK ROSE SYRUP

*Pour boiling water on a quantity of Damask roses just enough to cover them. Let them stand four-and-twenty hours. Then press off the liquor and add to it twice the quantity of sugar. Melt this and the syrup is completed.*

Thomas Tryon, *A Treatise of Cleanness in Meates,* 1692

▶ *Rosa* 'Albertine'

Roses have long been associated with English heraldry. *R. alba* was used by the House of York as its symbol, while the Red Rose of Lancaster was emblematic of that family during the Wars of the Roses. The 'Tudor Rose' used by Henry VIII bore red and white petals, indicating the melding of the two warring families. John Gerard, in his *Herball* (1597) referred to this when he wrote:

*Rosa* sp.

*. . . the rose deserved the chiefest and most principal place amongst all flowers whatsoever, being not only esteemed for its beauty, virtues and fragrant and odoriferous smell but also because it is the honour and ornament of our English sceptre . . .*

During the Middle Ages, roses were highly valued and were even used as rent payments. In 1576, the Bishop of Ely granted Sir Christopher Hatton permission to live in Ely House, on the sole proviso that the Bishop had free access to the garden and could harvest twenty bushels of roses each year.

The traditional housewife would not have grown roses just for their magnificent flowers and scent, however. The dried leaves are a good substitute for tea, the hips can be made into a delicious conserve or syrup, while the petals are useful as both a medicine and a cosmetic. Theophrastus' *Herball* was the first to mention the rose and medieval writers listed many of its uses; Peter Treveris, for example, recommended 'powdre of perles with sugre of roses' for heart complaints in his *Grete Herball* of 1526. Roses were advised for coughs, hayfever, baldness, as a children's tonic—even to facilitate hearing and memory! Queen Victoria is said to have favored a sauce made from Sweet Briar rose hips and lemon juice with her mutton. Her preference continued a long regal tradition of nibbling on roses . . .

### To Make a Cake with Rose Water the way of the Royal Princess, the Lady Elizabeth, daughter to King Charles the First

*Take half a peck of flowre, half a pinte of Rose-water, a pint of Ale Yeast, a pint of cream, boyl it, a pound and a half of Butter, six Eggs (leave out the whites), four pounds of Currants, one half pound of sugar, one Nutmeg, and a little salt, work it very well and let it stand half an hour by the fire, and then work it again, and then make it up, and let it stand an hour and a halfe in the Oven; let not your Oven be too hot.*

### To Candy Rose Leaves As Natural As If They Grow On Trees

*Take of your fairest Rose leaves [petals], Red or Damask, and on a sunshine day sprinkle them with Rose-water, lay them on one by one on a fair paper, then take some double refined Sugar beaten very fine, put in a fine lawne searse [sieve] when you have laid abroad all the rose leaves in the hottest of the sun, searse sugar thinly all over them and anon the sun will candie the Sugar; then turn the leaves and searse sugar on the other side, and turn them often in the sun, sometimes sprinkling Rose-water and sometimes searsing Sugar on them, until they be enough, and come to your liking, and, thus being done, you may keep them.*

William Rabisha, *The Whole Body of Cookery Dissected*, 1675

### To Make Conserve of Rose Hips

*Gather the hips before they grow soft, cut off the heads and stalks, slit them in halves, and take out all the seeds and white that is in them very clean; then put in an earthen pan, and stir them every day else they will grow mouldy; let them stand till they are soft enough to rub through a coarse hair sieve; as the pulp comes, take it off the sieve; they are a dry berry, and will require pains to rub it through; then add its weight in sugar, and mix it well together without boiling; keeping it in deep gallipots for use.*

E. Smith, *The Complete Housewife* (1736)

Rose vinegar, rose butter, rose honey, rose-hip marmalade, rose drops, rose juleps, electuary of roses, rosewater, pickled rosebuds, rose claret, rose syllabub and rose attar were just some of the many byproducts of these valuable flowers. Perhaps one of the most important is attar of roses, or rose oil, which is usually derived from *R. damascena*. A romantic legend has it that the Mogul prince, Jehangir, ordered roses to be floated in every canal running through the royal gardens to celebrate his wedding. His new wife, running her hands through the scented water, was fascinated to notice that a fragrant oil clung to her fingers and her doting husband ordered it bottled in her honor.

***Species:*** Although many of the modern varieties offer glorious color, I think that nothing can surpass the old-fashioned roses for fragrance and they should be the first chosen by anyone planning a scented garden: *Rosa alba*, described by John Gerard as having 'faire double flowers of a very sweete smell' and by Shakespeare as '. . . the milke-white rose with whose sweet smell the air shall be perfumed' (white, cream or pinkish large scented flowers with blue-gray foliage; 4–8ft/1.2–2.4m); *R. alba maxima*, or The Great Double White, cultivated in Bulgaria for precious attar (strongly scented white flowers; 6–7ft/1.8–2.1m). Among the *alba* roses, 'Belle d'Amour' and the fifteenth century 'Great Maiden's Blush' are wonderfully fragrant. *R. banksiae* is the well-known Banksia Rose with its daintily frilled blossoms (white or yellow flowers; vigorous climber with flowers appearing on old wood so *do not prune*); the 'Lutea' variety is the one most often seen, but *R. banksiae* 'Alba' is also worth seeking out as it is more fragrant. *R. bourboniana*, the beautiful, lushly-flowering Bourbon roses (white, pink, deep

*Rosa* 'Princess Margaret of England'

red and plum-colored blooms, are ideal border specimens; 4–5ft/1.2–1.5m); look for creamy-pink 'Mme Pierre Oger' or the stunning silvery-pink 'Zephirine Drouhin', a thornless variety thought to be Elizabeth I's 'rose without a Thorn'. *R. canina,* the old-fashioned Dog Rose, is still found growing wild in English hedgerows (pale pink single flowers; rampant grower with long, arching branches, so not suitable for a formal garden). *R. centifolia,* the deliciously scented Cabbage Rose, with masses of tightly curled petals, was once called the Hundred Leaves Rose (deep or pale pink flowers; 3–6ft/1–2m). 'Fantin Latour' is one of the best Centifolia roses and was a great favorite with the French painter it is named for. The dainty moss rose, *R. centifolia muscosa* (pink, crimson, plum to purple flowers, occasionally golden; 4ft/1.5m); the tea-scented China Rose, *R. chinensis* (pink, crimson and lilac blooms; small bushes 3ft/1m as well as large hedging types 8ft/2.4m, includes tiny 'fairy rose' varieties, which are ideal as pot plants). *R. damascena,* the Damask Rose referred to so often by Shakespeare, is possibly the most fragrant of all (double pink, perpetual flowering; 4ft/1.5m). Particularly lovely are the beautifully shaped flowers of 'Mme Hardy' and the rosy pink 'Omar Khayham', said to have descended from cuttings of the bushes grown on the rose-loving poet's grave. The Sweet Briar or Eglantine Rose, *R. eglanteria* (pink flowers; 8–9ft/2.4–2.7m), is a valuable hedging plant, smelling sweetest immediately after rain. Look for 'Janet's Pride' and 'Morning Blush'. *R. gallica,* the Apothecary's Rose, which was grown for its many medicinal uses, was also taken by the House of Lancaster as its symbol (pink, red, plummy-purple, and the famed 'Rosa Mundi' variety; 3–4ft/1–1.2m). The deep purple 'Duc d'Orleans' is also very lovely, as is the 'Sissinghurst Castle' variety. However, loveliest of all of the *gallica* roses must be the pink and white striped 'Rosa Mundi', named after the 'Fair Rosamund Clifford', mistress to Henry II. *R. moschata,* the Musk Rose (multi-petaled pink, white and yellow blooms); *R. rugosa* (large single dark red, white or pale pink flowers), possibly the hardiest of the old-fashioned roses that have long been cultivated for their glossy bright red hips, which were used to make syrup. 'Roseraie de l' Hay' has exquisite almond-scented deep crimson flowers while the popular 'Frau Dagmar Hastrup' has white semi double flowers followed by large red hips in autumn. *R. sempervirens* 'Princess Louise' (creamy flowers, richly scented with a rambling habit) and the *Wichuraiana* roses, *R. wichuraiana* from which many large-flowered rambling roses, notably 'Dorothy Perkins', are derived are also worth including in a scented garden—if you can ever find the space!

Among the newer roses, scented favorites include the hybrid teas: 'Ena Harkness', 'New Dawn', 'Sutters Gold', 'Crimson Glory' and 'Fragrant Cloud'. Of the Floribunda roses, the old favorite 'Ma Perkins' has been described as having a true 'wild rose' fragrance, while the aptly named 'Magenta' has a very soft, musky perfume. Gardeners with limited space should explore the dainty

### REFRESHING ROSE PERFUME

❖

*4 tbspns fresh red rose petals*
*1 cup vodka*
*1 tbspn fresh peppermint leaves*
*1 tbspn fresh rosemary leaves*
*1 tbspn grated orange zest*
*1 tspn grated lemon zest*
*2 cups boiling water*

*Cover rose petals with the vodka and store in a cool dry place for 7–10 days; strain and reserve the liquid. Bruise the peppermint and rosemary leaves, combine with the citrus zest and boiling water; allow to cool and strain. Pour both liquids into a lidded jar or bottle and shake thoroughly.*
*NOTE: keep rose perfume in refrigerator; once chilled, it is very refreshing on a hot summer's day.*

world of miniature or 'pony' roses, such as 'Baby Bunting' (double magenta blooms), 'Jackie' (yellow flowers), 'Pink Heather' and 'Sweet Fairy'.

*The rose looks fair, but fairer we it deem*
*For that sweet odour that doth in it live*

William Shakespeare, *Sonnets*

***Cultivation:*** Entire books have been devoted to this subject. The most important points to remember, I consider, are: mulching and regular composting, so plants are not subjected to drying winds; pay plenty of attention to picking, always cutting stems at a pruning angle just above outward-turning 'eyes' to encourage growth; rich soil and plenty of sunshine. In coastal areas, roses tend to fall victim to scale diseases, especially the nefarious 'black spot', so they should be planted where air may freely circulate around them rather than against a wall.

## *Rosmarinus*

ROSEMARY

*As for Rosemarine, I lett it runne all over my garden walls, not onlie because my bees love it, but because it is the herb sacred to remembrance and to friendship, whence a sprig of it hath a dumb language.*

Sir Thomas More

***Lamiaceae*** (Mint family).

***Scent:*** Pungent, tangy.

***Description:*** The beautiful aromatic rosemary with its glistening spiky gray leaves and pale blue flowers should hold foremost place among the fragrant leaves. Popularly known as 'the herb of remembrance', rosemary was worn by Greek scholars to help them recall their studies and it was also long used to represent fidelity. Its name means 'dew of the sea', a reference to its native haunts around the shores of the Mediterranean where

*Rosmarinus lavandulaceus.*

rosemary plantations once made the hillsides look as though they were cloaked in dew. Rather like lavender, rosemary has many medicinal, culinary and aromatic properties. Rosemary wine and rosemary cordial were much favored during Tudor times when it was also a popular topiary specimen. The chronicler Hentzner, in his *Travels* (1598), observed that rosemary featured in most English gardens. Interestingly, it was less likely to be a free-standing shrub than a wall specimen and at Hampton Court '. . . it was so plaited and nailed to the walls as to cover them completely.' Rosemary was carried at funerals and burnt in sick rooms, having the same sort of effect on the senses as smelling salts. It was much favored by the Countess of Hainault who wrote a long manuscript about it to her daughter, Queen Philippa, in 1370. Among other things, the Countess mentioned that crushed rosemary would 'gladden the spirits' and '. . . the leves layde under the heade whanne a man slepes, it doth away evell spirites . . .'

It was once customary to decorate the bodies of the dead with rosemary. Not only a favorite for solemn occasions, rosemary was seen at joyful ones as well, as when Robert Herrick instructed rosemary be grown '. . . for two ends, it matters not at all, Be't for my bridal or my buriall.' Bunches of gilded rosemary were exchanged by newlyweds and sprigs were worn by lovers attending New Year revels, inspiring Thomas Robinson's *Nosegay for Lovers* (1584):

*Rosemary is for remembrance*
*Between us day and night*
*Wishing that I might always have*
*You present in my sight*

SPIRIT OF ROSEMARY

*Gather a Pound and a half of the fresh tops of Rosemary, cut them into a Gallon of clean and fine Melasses Spirit, and let them stand all Night; next Day distill off five pints with a gentle Heat: this is of the nature of Hungary-Water, but not being so strong as that is usually made, it is better for taking inwardly: A Spoonful is a dose, and it is good against all nervous Complaints.*

From *The Receipt Book of Elizabeth Cleland* (1759)

Rosemary tea and wine flavored with rosemary have both been drunk since early times for their calming effect on the nerves. A distillation of the seeds and flowers 'drunke morning and evening, first and laste, will make the breath very sweete,' according to Thomas Newton's *A Butler's Recipe Book*. It is the principal ingredient in eau-de-cologne and other toiletry preparations and the leaves laid among linen are an effective moth deterrent. From a culinary point of view, rosemary is most often associated with lamb but, in fact, goes well with other meats too, especially pork. When next roasting a piece of meat, try mixing a little rosemary in the stuffing or add a few leaves to a marinade. Rosemary vinegar is a refreshing change of taste in summer salads and rosemary also goes well with fruit and fruit juice. An easy, aromatic honey may be made by putting a piece into a jar of clear honey.

***Species:*** *Rosmarinus* is a genus of two species, *R. officinalis* (white/mauve flowers; 5–6ft/1.5–1.8m) and *R. lavandulaceus*. There are a few different varieties offering interesting color and form, as well as the distinctive pungent aroma; *R. o.* 'Miss Jessop', which is a more shrubby bush of upright habit, is very suitable as hedging, bearing lavender colored flowers; *R. o.* 'Prostratus' and *R. lavandulaceus*, which both hug the ground, will creep most attractively over walls and banks.

***Cultivation:*** Rosemary prefers a warm, sheltered position, ideally receiving full morning or afternoon sun. A well-drained sandy soil will produce the best results and an occasional dressing with lime will not go astray. Rosemary is hardy and can be readily propagated from seeds or cuttings; the bushes should be trimmed every autumn. Pinching out new growth during summer will produce a thicker plant.

ROSEMARY MOTH SACHETS

Small sachets of dried herbs have been used to scent linen and clothes for centuries; they also kept away moths and fleas. Richard Banckes wrote of rosemary in his *Herball* of 1525: 'Also take the flowres and put them in a chest amongst your clothes or amonge bookes and moughtes [moths] shall not hurte them . . .' A sachet to lay among linen:

❖

*2oz/50g rosemary*
*1oz/25g hyssop*
*1oz/25g wormwood*
*1oz/25g lavender*
*1–1½ tbspns orris root powder*

❖ ❖ ❖

## *Ruta*

RUE

*There's rue for you; and here's some for me:*
*We may call it Herb O'Grace o'Sundays:*
*O, you must wear your rue with a difference . . .*

William Shakespeare, *A Winter's Tale*

***Rutaceae*** (Rue family).

***Scent:*** Sharp, pungent.

*Ruta graveolens*

***Description:*** Rue was also known as the Herb of Grace or herb of repentance because of its extremely bitter taste. In *A Winter's Tale,* Perdita says:

*For you there's Rosemary and Rue: . . .*
*Grace and remembrance be to you both . . .*

Rue is a most decorative herb with strongly scented perennial blue-gray leaves and small yellow flowers in summer. Although pungent, rue leaves are most refreshing to inhale when they are crushed. They have a long history of medicinal use and Milton mentioned its use for eye ailments:

*Michael from Adam's eye the filme removed . . . then purged with Euphrasie and Rue the visual nerve, for he had much to see . . .*

Along with lavender, rue was a chief constituent of the famed 'Four Thieves Vinegar', said to have been used by robbers during the Great Plague of Marseilles: they washed their hands and faces with the powerfully scented vinegar before stealing from the homes of ill folk and never caught the disease themselves. Culpeper wrote that the leaves were used to strew the floors of prisons and workhouses to reduce the risk of infection and the ninth century monk, Walfred Strabo, noted that 'great is its power over evil odours'.

***Species:*** *Ruta graveolens,* the Common Rue, has the most attractive serrated leaves (blue green foliage; pale yellow flowers; 2ft/60cm), which have a whiff of orange about them. The *R. graveolens* 'Jackman's Blue' has almost pure blue leaves with a pleasing aromatic perfume. Avoid *R. chalepensis;* although attractive to look at, it has a quite rotten, fetid smell, especially after rain.

THE STING OF A BEE OR A WASPE

*Take Rue a handfull and stampe the Juyce of the leaves and aply to any part hurt by the Sting of a bee or a waspe. A present Remedy.*

*M.S. Book of Receipts,* by Thomas Newington, 1719

***Cultivation:*** Rue is shallow-rooted and likes a sunny aspect. It will do well in quite poor soil. A dry, sandy loam with some mixed-in chalk is best of all. Rue may be propagated from seed sown in summer or cuttings and should be clipped each spring, though never into the old wood. Trim off new growth instead and use as cuttings.

## *Sambucus*

ELDER

### To Take Away the Freckles in the Face

*Wash your face, in the wane of the Moone, with a sponge, morning and evening, with the distilled water of Elder-leaves, letting the same dry into the skinne. Your Water must be distilled in May. This was from the Traveller, who hath cured himself there by.*

Sir Hugh Platt, *Delightes for Ladies*, 1659

***Capricoliaceae*** (Honeysuckle family).

***Scent:*** Musky and sweet.

***Description:*** Elder trees are found throughout the world, especially in gardens of the superstitious, for all herbs are thought to be under the protection of the Spirit of the Elder. The Common or Black Elder, *Sambucus nigra*, was believed to ward off lightning and witches and one is usually found near very old buildings, for example, there is an ancient elder tree just outside Westminster Abbey.

French fruit growers always store their pear crops in elderflowers, so the muscatel perfume may be imparted to the fruit.

The wood is very hard but easily carved and is widely used to make musical instruments; 'Pipe Tree' is an old country name for the elder. The tree also has many healing properties: water distilled from the flowers and leaves is a valuable and aromatic astringent; a poultice made from the berries will ease sunburn and elderflower cream (made by macerating the flowers in beeswax) is

### The Lady Thornburgh's Syrup of Elders

*Take Elderberries, when they are red, bruise them in a stone Mortar, strain the juyce, and boyl it to a Consumption of almost half, skum it very clear, take it off the fire whilest it is hot, put in Sugar to the Thicknesse of a Syrup; put it no more on the fire, when it is cold, put it into glasses, not filling them to the top, for it will work like Beer.*

*The Queen's Closet Opened*, by W. M., Cook to Queen Henrietta Maria, 1655

*Sambucus racemosa*

### Elder Flower Nourishing Night Cream

This is an excellent regenerative treatment and particularly appropriate for dry or mature skin. Lavish it on elbows, knees or roughened hands to soften and smooth. Also very good on sunburnt shoulders and noses.

❖

*2 tbspns dried elder blossoms*
*1½fl oz/50ml buttermilk*
*2 tbspns beeswax*
*2 tbspns lanolin*
*2 tbspns olive oil*
*1 tspn almond oil*

*Combine the elder flowers and buttermilk over low heat for 30 minutes. Remove from heat and strain. Place in double boiler and slowly combine with the beeswax, lanolin and oils, stirring continuously until all are melted. If desired, a few drops of honeysuckle perfume may be added. Pour into sterile, warmed glass pots; cool and cap securely.*

a marvelous complexion aid. Both elderflower vinegar and 'rob', or syrup, are time-honored remedies for colds and coughs.

The elder was once thought unlucky, with Sir John Mandeville averring it was 'the tree of eldre that Judas henge himself upon'. Many still believe it was. Others say an elder tree will bring fairies to the bottom of your garden and even the normally pragmatic Eleanour Sinclair Rohde assures us that 'If you . . . stand under it at midnight on Midsummer Eve . . . then you will see the King of the Elves go by.'

***Species:*** *Sambucus nigra,* the Common Elder (cream colored blossoms; 20ft/6m), is pleasant to behold and to smell. The flowers have an appealing, musky scent. Also attractive are *S. canadensis* (white flowers; 12ft/3.6m) and *S. racemosa* 'Plumosa Aurea', but avoid *S. ebulus,* whose lovely pink and white flowers, unfortunately, have an acrid, foxy smell.

***Cultivation:*** Elders grow throughout the world and will do well in both open or shady situations. A tree requires plenty of sun in order to flower and a sheltered, still aspect will maximize the enjoyment of the scent in a garden.

## *Santalum*

Sandalwood

***Santalaceae*** (Sandalwood family).

***Scent:*** Exotic, sharp, aromatic.

***Description:*** Sandalwood is a small parasitic tree native to the forests of Malaysia, Timor and parts of India. It was mentioned in early Sanskrit literature and the wood and essential oil have both

### Oriental Potpourri

A sweet, soft potpourri for the bedroom or parlor.

*2oz/50g rose petals*
*2oz/50g sweet jasmine flowers*
*1oz/25g orange flowers*
*1oz/25g basil*
*1oz/30g sandalwood chips*
*2 tspns coriander seeds*
*2 tspns crushed cumin seeds*
*2 tbspns gum benzoin powder*
*4–6 drops jasmine oil*

*Dry all the flowers and leaves until they are crisp (either in a slow oven or somewhere sunny) and place them in a glazed crock. Add the sandalwood chips, coriander, cumin and gum benzoin. Using your hands, mix well. Add the oil, one drop at a time, and mix well after each addition.*

been used in embalming and funeral rites for Singhalese princes, and for burning as an incense. The Indians also regarded sandalwood, or 'sanderswood', as a valuable medicine and would mix the oil with rice gruel as a cooling astringent treatment for malaria and other feverish ailments. However, its most popular use by far is in perfumery, either as an essence in its own right, as a complementary fragrance in most rose perfume types, or as a fixative. Sandalwood powder or scrapings are included in many old potpourri recipes.

Sandalwood chips are highly repellent to insects and may be scattered among clothing to deter moths and this is why clothes boxes and presses were often made from sandalwood. The intense harvesting of sandalwood for such purposes, however, has meant that many forests are now under government protection.

***Species:*** *Santalum album* (40ft/12m) is the variety cultivated for its scented wood, roots and oil; *S. rubrum*, or Red Sandalwood, is used for coloring and dyeing.

***Cultivation:*** Sandalwood is a tropical tree and would be suitable in a large, exotic and rather jungly garden.

## *Santolina*

LAVENDER COTTON

***Asteraceae*** (Sunflower family).

***Scent:*** Strong, pungent.

***Description:*** Although it is sometimes also called French Lavender, Lavender Cotton is not a true lavender at all. It is a very beautiful perennial plant with yellow clusters of flowers and feathery silver-gray toothed foliage. The leaves were once used as a children's vermifuge and the Arabs are said to use the juice from the stem for bathing sore eyes. Culpeper wrote that Lavender Cotton 'resisteth poison, putrefaction and heals the biting of venomous beasts'.

Lavender Cotton is now mainly used as a low hedge or edging for borders, its pretty leaves creating a silvery carpeting effect. In much the same manner as lavender, the dried twigs and leaves of Lavender Cotton may be placed among clothes to deter moths. The French called it *garde-robe* and mixed it with southernwood and lavender as a 'swete bag' mixture for the linen press.

***Species:*** Prettiest is *Santolina chamaecyparissus* (yellow flowers; 3-4ft/1-1.2m). *S. neapolitana* (bright green flowers; 3ft/1m) is well known but, be warned, some find the pungent odor from the flowers rather unpleasant. But others find this pungency refreshing and quite a pleasant contrast to honey-type scents in the garden.

***Cultivation:*** Lavender Cotton is a native of Mediterranean regions and, as such, will do well with plenty of sun and a well drained soil. Cuttings should be taken in spring and will strike quickly, being ready for transplanting before winter. Lavender Cotton should be clipped quite hard in autumn, after flowering, to ensure plants remain thick and dense, especially if they form a hedge.

## *Saponaria*

SOAPWORT

***Caryophyllaceae***

***Scent:*** Sweet, with a slight hint of cloves.

***Description:*** The Common Soapwort (*Saponaria officinalis*) is a perennial plant native to Britain and parts of Central Europe where it may be seen growing wild in hedgerows. With its pretty carnation-like double pink or white flowers, it is popular as a cottage garden plant. Soapwort was first used by the Romans, who called it *struthinin*. They found that by crushing the leaves, a mucilaginous juice that lathered in hot water and removed grease and stains, was released. The Anglo-Saxons mixed it with willow ash and then simmered it in rain water. Medieval monks called

▶ Delicate blue forget-me-nots are combined with a spring bulb border.

*Saponaria officinalis*

it Fuller's Herb and other old country names, such as Hedge Pink, Latherwort, Wild Sweet William, Bouncing Bet and Bruisewort. To Gerard's assertion that soapwort's leaves 'yielded a juice which when bruised, scoureth almost as well as soap', Mrs Leyel adds that 'a decoction cures the itch' and 'for old venereal complaints it is a good cure'! Certainly it has disinfectant properties and a leaf placed over a cut finger or grazed knee is a useful first aid tip to remember.

***Species:*** The main species of saponaria for cultivation in a scented garden is *S. officinalis*. The double pink (*S. officinalis* 'Rubra Plena') and white forms (*S. officinalis* 'Alba Plena') are charming in an older style garden, too. They will grow to at least 2ft/60cm and I have seen ones with 4ft/1.2m stems!

***Cultivation:*** Soapwort is a sturdy perennial, requiring semi shade and plenty of water in order to flower in summer. The attractive smooth and slightly glossy leaves mean soapwort is a handsome bedding plant year-round.

### SOAPWORT WASHING LATHER

It is easy to prepare a washing lather from soapwort. Its mild scent and disinfectant properties make this an ideal additive for children's baths.

❖

*Put 1½oz/40g of soapwort leaves and roots into a saucepan and cover with 2 cups of rain or spring water. Gradually bring to the boil and simmer for 3–5 minutes. Cover and allow to cool, then strain the liquid through a non-aluminum colander, pressing down well on the soapwort mixture. Decant into a sterile jar or bottle and cap securely. Use about ½ cup per bath.*

## ***Satureja***

SAVORY

*Here's flowers for you;*
*Hot Lavender, Mints, Savory, Marjoram . . .*

Wm Shakespeare: *A Winter's Tale* (IV iv)

***Lamiaceae*** (Mint family).

***Scent:*** Pungent, refreshing.

***Description:*** The botanical name of this plant indicates that it was thought to be a 'plant of the satyrs' by the Romans. They considered it an aphrodisiac, a piece of folklore that endured for some centuries with Richard Banckes writing in 1525 that:

*Satureja montana*

*It is forbidden to use it much in meats . . .*
*[since] . . . it stirreth him that useth lechery*
*[but if drunk in wine would] make thee a*
*good meek stomach.*

Virgil recommended that savory be planted close to bee hives—not only do the inhabitants dote on the nectar-rich flower but the pain of a bee or wasp sting is rapidly relieved by rubbing it with crushed savory leaves.

Winter Savory was a traditional strewing herb and may be used in potpourri or posies to good effect. The strongly aromatic leaves make it a must for any scented garden. In *The Arte of Gardenynge* (1563), Thomas Hyll wrote that 'knotte gardens' were usually 'sette with Isope and Thyme or with winter Savory and Thyme, for these endure all the winter through green.'

Savory leaves have long been valued for their hot, spicy flavor. The Romans made a sauce from them by adding vinegar (not unlike today's mint sauce), which they would pour over roasted meat. In early medieval times, it was mixed with breadcrumbs and used as a coating for meat or fish, and also as a stuffing for poultry. One fourteenth-century cook wrote:

*. . . take sage, parsley, hyssop and savory, quinces*
*and pears, garlic and grapes, and fill the geese*
*therewith and sew the hole . . .*

Used sparingly, savory adds zest to green salads and egg dishes. It is particularly useful when cooking all varieties of peas and beans, not only adding flavor, but also reducing cooking smells.

***Species:*** *Satureja montana*, Winter Savory, is a small woody perennial shrub; *S. hortensis*, Summer Savory, is a half hardy annual with slightly pinkish leaves that are larger than those of Winter Savory. Summer Savory was reputed to preserve the sight. Gerard named it St Julian's Herb because he first saw it growing wild on St Julian's Rocks by the Tyrrhenian Sea in Italy.

***Cultivation:*** Both Summer and Winter Savory may be propagated from seeds sown in spring. Winter Savory is easily increased by taking cuttings of the young shoots, with a 'heel' attached, in spring, or by root division in autumn. Always cut the plants back after flowering. They will become woody and less leafy next year if this is not done. Winter Savory is extremely hardy, preferring full sun and a poor, pebbly soil; it will even retain its leaves under a covering of snow. It is very useful as a border or edging plant or for a rockery. If you wish to use the leaves of either variety for drying, harvest them in midsummer and again in early autumn.

## *Scilla*

SQUILL, ENGLISH BLUEBELL

***Liliaceae*** (Lily family).

***Scent:*** These bulbous plants were named the 'starry jacinths' by Parkinson and they are, indeed, like dainty little stars. Scillas are native to England and Europe, growing as far north as Siberia, where those with the patience to pick a posy of *Scilla siberica*, the Siberian Squill, will be rewarded by a particularly intense fragrance. In the wild, scillas grow in great drifts beneath shady trees or in

*Scilla* sp.

meadows. They will prefer similar conditions if you want to naturalize them in a garden, so plan to mass them beneath trees or underplant shady hedges with the bulbs.

***Species:*** *Scilla bifolia* (bright blue flowers; 6in/15cm); *S. hyacinthoides* (pale mauve/blue flowers; 18in/45cm); and *S. sibirica* (blue flowers; 4in/10cm), which has blue pollen much sought after by bees. 'Spring Beauty' is the *sibirica* variety to look out for; it has deep blue flowers of matchless beauty and scent. *Endymion (scilla) non-scriptus* the English native bluebell, is sometimes included in catalog listings. It has deep blue flowers on long stems (1–1½ft/30–45cm).

***Cultivation:*** Scillas are extremely hardy, only requiring plenty of moisture and semi shade. They will increase rapidly as the bulbs self-corm and spread through the garden. They are also ideal for planting in a dish and bringing indoors, where the delicate perfume will drift through the house.

## *Selenicereus*

QUEEN OF THE NIGHT

***Cactaceae*** (Cactus family).

***Scent:*** Exotic, powerful, rich, vanilla-like.

***Description:*** This is a group of night-flowering cacti native to Mexico and the West Indies and they are one of the wonders of nature. They take their name from the Greek *selene,* meaning moon, for this is the time when the flowers emit their powerful scent. Be warned: by day, these plants have little to recommend them. They are

*Selenicereus grandiflorus*

straggling climbers with aerial roots and (very) sharp needle-like spines, so they should be planted out of sight and out of reach of children. However, at dusk the great tubular flowers open (8in/20cm wide), white inside and yellow without, and give off mighty 'puffs' of rich perfume for a considerable distance. These 'puffs' occur because the flower's calyx makes a spasmodic movement, and they are part of nature's plan for maximizing the flower's chances of a visit from the night-flying moth, Lepidoptera. By dawn, the flower will be spent; a short, but certainly glamorous, life.

***Species:*** *Selenicereus grandiflorus* and *S. grandiflorus* 'Mexicanus', the popular Mexican Queen of the Night (white and yellow flowers). The height of these plants vary, but they are vigorous and will certainly grow quite large; old, well-established plants have stems up to 30ft/10m and, with correct conditions, will still add an extra 6ft/1.8m in a single year.

***Cultivation:*** In tropical warm areas, selenicereus plants will grow best in a rich soil against a sun-drenched wall. Though the flowers are quite wonderful to gaze upon at night, the plant is not that attractive during the day so either interplant with another more decorative trailing plant, such as bougainvillea, or train under a balcony or over a pergola out of the main garden area. It requires a minimum winter temperature of 50°F/10°C so, in cooler climates, should be afforded greenhouse cultivation. A generous mulching with leaf mold and perhaps a little lime during spring will enhance summer flowering.

## *Skimmia*

***Rutaceae*** (Rue family).

***Scent:*** Powerful, lingering, similar to lily of the valley.

***Description:*** Skimmias are small evergreen free-flowering shrubs with deep green tough leathery leaves, which are relatively unharmed by smoke or pollution, making them a good aromatic choice for an inner city garden or a position close to traffic. The leaves have a spotted appearance, because they are dotted with oil glands that release a pungent aroma when crushed. (NOTE: the scent of the leaves is not always as pleasant as the pronounced sweet perfume of the flowers.) The small white flowers are replaced by dainty clusters of reddish-coral berries, which often remain for months, adding color to wintry days.

***Species:*** *Skimmia japonica* (white flowers; 4ft/1.2m) is one of the most popular shrubs, but because the male and female flowers are borne on separate plants, you need to plant at least one male and two females to ensure that fertilization occurs. Interestingly, it is the male flowers that produce the most perfume, while *S. japonica* 'Fragrans' provides the most intense perfume. For a smaller garden where only one shrub will really fit, plant the Chinese native *S. japonica* subsp. *reevesiana* (greenish white flowers; 3-4ft/1-1.2m). Unlike *S. japonica,* this species has bisexual flowers. It also provides the cheerful display of rosy berries, which is the skimmias' main attraction. *S. laureola,* from the Himalayas, is less hardy (yellow flowers; 4-5ft/1.2-1.5m), with red berries and heavily scented foliage. Its flowers are intensely scented, so much so that they may be considered oppressive if planted close to the house. However, the fragrance is very enjoyable at a little distance, so plant *S. laureola* in a warm, shady corner by the back fence.

***Cultivation:*** Skimmias will do well in most cool, temperate areas. They prefer partial shade and a moist, rich, acid soil with plenty of leaf mold. When they have finished flowering, cuttings taken from skimmias will usually do well; root them in a sandy type of potting mixture for best results.

*Stephanotis floribunda*

## *Stephanotis*

***Asclepiadaceae***

***Scent:*** Powerful, rich.

***Description:*** Within this group of twining plants it is *Stephanotis floribunda*, a beautiful evergreen, which is the most fragrant. Hailing from Madagascar, where it is variously known as Clustered Wax Flower, Chaplet Flower and Madagascar Jasmine, it is free flowering and produces masses of waxy, white, bouvardia-like flowers, which are deliciously fragrant. As its country name suggests, stephanotis has long been a popular choice for bridal wreaths and bouquets. The Victorians were much enamored of it, and devoted much of their leisure time to training the vines up wires and across the roofs of their large glass greenhouses and conservatories. It was an expensive cut flower in England because large amounts of fuel were required for its cultivation under glass.

***Species:*** *Stephanotis floribunda* (pure white flowers with a waxy look; twining stems up to 30ft/10m long).

***Cultivation:*** Stephanotis does well in warm areas, adapting rapidly both to full sun or warm greenhouse conditions. It requires a minimum winter temperature of 60°F/15°C. The important thing to remember about stephanotis is that it resents root disturbance. If you allow plenty of room for root expansion in a rich loamy soil and provide more water in summer, when it is growing, than in winter, your stephanotis should be a glorious sight.

### SCENTED HERB AND HONEY DRESSING

A light, smooth dressing, perfect for a crisp green salad.

❖

*1 tbspn fresh marjoram*
*2 tbspns fresh green parsley*
*2 tbspns fresh rosemary*
*2oz/50g raw honey*
*4 cloves garlic, crushed*
*5fl oz/150ml apple cider vinegar or white wine vinegar*
*5fl oz/150ml olive oil*
*pinch curry powder*

*Blend all the ingredients, except the olive oil, at medium speed in a food processor. Stir the olive oil carefully through the mixture, then shake well just before serving.*

## *Syringa*

LILAC

*I remember, I remember*
*The roses red and white,*
*The violets and the lily-cups—*
*Those flowers made of light!*
*The lilacs where the robin built,*
*And where my brother set*
*The laburnum on his birthday—*
*The tree is living yet!*

Thomas Hood, *Past and Present* (1799–1845)

***Oleaceae*** (Olive family).

***Scent:*** Sweet and penetrating.

***Description:*** Syringa is the botanical name for lilac, the harbinger of summer, whose fragrance will fill a garden with sweetness. 'Lilac' is a Persian word for 'flower', which accompanied the plant's introduction to Europe and England and has remained with it to this day. Early plantings of lilacs were recorded in Henry VIII's day. He requested that they be set in a circle around a fountain in his garden at Nonsuch and a contemporary historian grudgingly acknowledged these '. . . six lilac trees . . . bear no fruit but only a pleasant smell'.

Lilacs are deciduous bushy shrubs with smooth heart-shaped leaves and heavy pyramidal clusters of double or single tubular flowers. Depending on the variety, these flowers may be white, pink or deep purple; the white flowers have the most delicate scent and the darker ones are more powerfully fragrant. The scent of lilacs is especially pronounced after rain. Also, since the perfume glands are in the flower petals, it stands to reason that double flowered varieties will have a stronger perfume.

***Species:*** *Syringa vulgaris*, the Common Lilac (mauve-purple flowers; 9ft/2.7m) is well known, along with many cultivars including 'Clark's Giant' (deep blue flowers) 'Edmond Boissier' (rich purple flowers), 'Primrose' (pale yellow) and 'Mme Buchner' (plummy rose). Rouen lilac, *S.* x. *chinensis* (deep mauve blooms; 9ft/2.7m), is a very old variety, dating back to 1777. Other species suitable for mixed shrub borders are the Persian Lilac, *S. persica* (pale blue or mauve flowers; 7ft/2m).

***Cultivation:*** Lilacs are hardy in almost all cold or hilly districts and require a cold winter to bring on flowering in the spring. Ordinary, well-drained soil is quite adequate, provided a good mulching is given each year. Lilacs do have a rather sparse, twiggy habit, but this must be tolerated and the plant only very lightly pruned as flowers only appear on new growth. Plant lilacs in a cool shady place in the garden. Propagate by cuttings rooted in sandy soil.

## *Tanacetum*

TANSY

*Fragrant the Tansy breathing from the meadows*
*As the west wind blows down the long green grass;*
*Now dark, now golden, as the floating shadows*
*Of the light clouds pass, as they were wont to pass*
*A long while ago!*

John Clare (1793–1864)

***Asteraceae*** (Sunflower family).

***Scent:*** Pleasantly pungent and also slightly camphoraceous.

***Description:*** Tansy is a pretty little herb with very beautiful dark green fern-like leaves and bright yellow round flowers, which explain the derivation of its common name, 'Golden Buttons'. The name 'tansy' comes from the Greek '*athanasia*', meaning immortality, and was bestowed upon this plant because of its extreme hardiness (be warned: it is irrepressible in a garden!) and because eating the leaves or drinking tansy tea were both once thought to confer longevity.

Tansy has been cultivated since Saxon times for its many culinary and medicinal uses, primarily in stuffings or for seasoning meat. Before mint became popular as an accompaniment to roast lamb, tansy sauce was served. It was used to flavor cakes eaten at Easter, as evidenced by the old song:

*Soon at Easter cometh Alleluyah,*
*With butter, cheese and tansy . . .*

Tansy puddings were commonly eaten during spring, tansy wine is an old remedy for stomach troubles and tansy tea was recognized as a cure for rheumatism. There was also a traditional dish known as 'tansy', which was a type of omelet that could be adjusted as sweet or savory. Try including chopped tansy leaves in egg dishes or mixing them with peppercorns and sea salt, then rubbing the mixture over a loin of pork before roasting it . . . delicious! For the more adventurous, why not prepare a traditional 'tansy'? As this old recipe shows, this dish was a meal in itself:

A GOOD TANSY

*Take seven eggs and leaving out two whites, and a pint of Cream some Tansy, Thyme, Sweet Marjoram, Parsley, Strawberry leaves all, shred very small a little nutmeg, add a plate of grated white bread, let these be mixed all together, then fry them but not too brown.*

From *The Receipt Book of John Nott,* Cook to the Duke of Bolton, 1723

In the north of England, such a tansy cake was a popular forfeit offered by young lads and lasses who were flirting over a game of hide-and-seek or handball. Robert Herrick mentions this:

*At stoolball, Lucia, let us play,*
*For sugar cakes or wine;*
*Or for a tansy let us pay,*
*The loss be thine or mine . . .*

Tansy's antiseptic and insect-repelling properties make it very useful in the kitchen. It was once used widely as a strewing herb and great bunches hung in larders or meat safes for the same reason. Gather and dry the leaves and flowers and

*Tanacetum vulgare*

add to a spicy kitchen potpourri containing marigold flowers, nutmeg chips, southernwood and perhaps a drop or two of cedarwood oil. Not only will such a mixture have a lovely sunshiny appearance, it will deter flies, lice and fleas.

***Species:*** *Tanacetum vulgare,* the Common Tansy (bright golden yellow, button-shaped flowers; 2ft/60cm).

***Cultivation:*** Tansy is extremely hardy and easy to grow, however, you have to be absolutely ruthless to keep it in check! Propagate tansy from seed and keep it well watered during spring and summer, the leaves may be harvested in spring and the plants should be cut back in autumn. Ideally, tansy should be 'contained', for example, in a kitchen window box, or a bricked-in herb garden. If not, the long tangled roots will invade the rest of the garden.

### INVIGORATING TANSY SCRUB

Simple to make, this facial scrub will bring a healthy glow to sallow skin. Tansy has an enlivening effect on skin and mildly bleaching properties, which help reduce inflammation.

*3 tbspns tansy flowers, crushed*
*2–3 tspns fresh parsley*
*2oz/50g ground almond meal*
*2oz/50g Fuller's Earth*
*5fl oz/150ml water*

*Infuse the tansy and parsley in boiling water for 30 minutes. Cool, strain and mix the liquid with the almond meal and Fuller's Earth to form a gritty paste. Apply to slightly damp skin with light, circular movements. Rinse with lukewarm water and pat dry with a soft towel.*

## *Thymus*

THYME

*The opening summer, the sky,*
*The shining moorland—to hear*
*The drowsy bee, as of old,*
*Hum o'er the thyme . . .*

Matthew Arnold (1822–1888)

***Lamiaceae*** (Mint family).

***Scent:*** Strong and fresh, pungent.

***Description:*** Tennyson wrote of 'the thymy plots of Paradise' and, indeed, no scented garden should be without at least one of the many varieties of thyme. They are native to the Mediterranean and warm areas of Europe and are all extremely fragrant. Their tiny leaves and pink or purple flower heads make them a very attractive choice for the front of a herb garden or the edge of a border. With their creeping habit, some thymes, for example *Thymus serpyllum,* may be planted to form a fragrant carpet or 'lawn'. Francis Bacon referred to this idea when he wrote of '. . . those flowers which perfume the air most delightfully . . . being trodden upon and crushed, Burnet, Wild Thyme and Water Mint.' The thymes may also be used about a rockery or set between paving stones. Other thymes, notably *T.* x *citriodorus,* have a compact bushy habit; they make attractive small shrubs or even topiary specimens.

The ancient Greeks burnt thyme at the altars to their gods as well as using it medicinally and in the kitchen. To say that anyone 'smelt of thyme' was a fine compliment, expressing great praise for that person's appearance. The Romans used thyme as a remedy for melancholy and to flavor cheese and liqueurs. It was frequently used in Lancastrian England, when ladies used to give a sprig of thyme to their knights so that they might be courageous in jousting tournaments. They would also often embroider their lover's scarf with a bee hovering over a sprig of thyme, a reference to the great activity of bees about thyme flowers. During these times, when honey was one of the few sweeteners known to man, those plants visited by bees were considered very important. Thomas Hyll, in *The Gardener's Labyrinth* (1577) wrote that '. . . the owners of Hives have a perfite foresight and knowledge what the increase or yeelde of Honey will bee everie yeare by the plentifull or small number of flowers growing and appearing in the thyme about the summer solstice.' Not only do bees love thyme, it is also said to be a favorite plant of fairies . . .

### To enable One to See the Fairies

*A pint of sallet Oyle and put it into a vial glasse; and first wash it with rose-water and marygold water; the flowers to be gathered towards the east. Wash it till the oyl becomes white, then put it into the glasse, and then put thereto the budds of hollyhocke, the flowers of marygolde, the flowers or toppes of whild thyme, the budds of young hazle, and the thyme must be gathered near the side of a hill where fairies use to be; and take the grasse of a fairy throne; then all these put into the oyle in the glasse and sette it to dissolve three days in the sunne and then keep it for their use.*

Receipt dated 1600, Ashmolean Museum, Oxford

### THYME-SCENTED CANDLES

❖

*12½oz/350g paraffin wax*
*1oz/35g stearin*
*pink or yellow candle dye powder (from hobby shop), if desired*
*few drops honeysuckle essential oil*
*1oz/25g dried thyme leaves*

*Prepare the molds. Melt the wax in a double boiler over a low heat. Melt the stearin and dye in another double boiler, then add oil. Stir thyme into the melted wax, then stir in the stearin mixture. Mold candles and trim wicks according to the manufacturer's instructions. Polish candles with cotton wool that has been dipped in essential oil.*

Common thyme (*Thymus vulgaris*) is the one most used for cookery. It is an essential part of *bouquet garni* and also goes well with most meats and egg dishes; it is widely employed in vegetarian cookery as it marries well with nuts and pulses. Lemon, apple and orange thymes can also add their special flavors to food; lemon thyme complements fish and poultry, orange thyme is lovely and refreshing in a fruit salad. Caraway thyme (*T. herba-barona*) was traditionally used to preserve meat and was the accompanying relish for a 'baron' of beef, hence the derivation of its formal name. Despite their many culinary uses, thyme is most often grown for its fragrance and pretty appearance. Perhaps things have not changed so much from when John Parkinson wrote in 1629:

*To set down all the particular uses whereunto thyme is applyed were to weary both the writer and the reader . . . we preserve them with all the care we can in our gardens for the sweete and pleasant sents they yield . . .*

In ancient Rome, thyme was burnt in homes and public places as a fumigatory practice, which also drove away fleas and other vermin. Dried sprigs may be laid among furs and winter clothes, in much the same way as lavender or bay leaves, to deter moths. The many scented varieties of thyme may be used in sweete sachets or potpourri as well as in aromatic tea, which is 'a great refreshment to the fading or drooping spirits', according to Mrs Leyel. The dried leaves may be used to make perfumed candles.

***Species:*** *Thymus azoricus* (bright green leaves, powerful orange scent; prostrate); *T.* x *citriodorus*, Lemon Scented Thyme (lemony scent; 9in/25cm); *T. serpyllum coccineus* (rich pink flowers and dark leaves; prostrate); *T. herba-barona*, Caraway Thyme (rose pink flowers); *T. serpyllum* 'Albus' (white flowers; prostrate habit), and *T. vulgaris*, Common Thyme (gray-green foliage; 9in/25cm).

***Cultivation:*** In an old description of thyme, one reads '. . . they joy to be placed in an open and sunny space'. Folkloric wisdom adds thyme will only flourish where the air is pure but it is reasonably hardy, so do not despair if you have a city garden. Thyme prefers a dry, well-drained soil and full sun. It may be propagated by sowing seed in the spring or taking cuttings in summer with a 'heel' attached and rooting them in potting mix. To make a traditional thyme 'alleye' alongside a path, as described by Francis Bacon, set groups of 3 or 4 plants about 10in/25cm apart and they will grow together to form a firm carpet. If the soil has been thoroughly cleared of perennial weeds and if the lawn is hand-weeded during its first season, the mat-like foliage should mean very little weeding will be necessary in the future.

### SCENTED TEAS

Scented flower teas are calming and refreshing, as they do not contain the caffeine present in ordinary tea or coffee. In summer, chill scented flower tea and add a little honey for a delicious iced drink. Do not forget that picking the flowers for your teapot is itself a relaxing pastime. Experiment with the flowers—the following combinations are very pleasant:

❖

*lavender, chamomile and bergamot*
*wild rose petal and chamomile*
*cowslip*
*lavender and rosemary flower*
*lime flower and mint*

## Verbena

***Verbenaceae*** (Vervain family).

***Scent:*** Soft and very sweet.

***Description:*** Verbenas are small perennials that are native to South America, and so tend to prefer a warm, dryish climate. With their superb array of color, sweet perfume and semi-trailing habit, verbenas are a delightful choice for a hanging basket as well as for massing in bold groups at the front of flower beds or along paths. For a sensational 'olde-worlde' effect, plant an old-fashioned rose, like 'Souvenir de la Malmaison', with pink verbena and blue rue for a bouquet of spring and summer fragrance!

Wild verbena or vervain (*Verbena officinalis*) was once thought to have mystic qualities. The Greeks said it was used to clean Jupiter's table and plate before he was served food and Venus is usually depicted wearing a crown of thyme and verbena flowers. The druids offered sacrifices to the earth before cutting a single stem of verbena and, later, its sweet perfume and mystic prowess meant it was a popular ingredient in love potions.

*Verbena peruviana*

### LEMON VERBENA SACHET BAGS

❖

*2oz/50g lemon verbena*
*1oz/25g marjoram*
*dried zest of 1 lemon, crushed*
*1 tbspn cloves, crushed*
*1 tbspn orris root powder*

*Dry all the herbs until they are crisp and finely crumble them. Make small square or heart-shaped sachets from natural fibers, such as cotton or silk, and fill loosely with the mixture before tying closed with ribbon or a piece of lace.*

❖ ❖ ❖

For centuries dried verbena, especially the lemon-scented variety, has been packed in small sachets for storing in the linen cupboard.

***Species:*** *V.* x *hybrida grandiflora* (showy clusters of red, pink, yellow, purple, blue or white flowers; 12–18in/30–45cm) is most attractive. *V. peruviana* has brilliant scarlet flowers and a spicy scent.

***Cultivation:*** Verbenas are half hardy in cool temperate climates, provided the soil is moist and it receives full sun. They may be grown from seed set early in spring with the seedlings set out in the garden in summer where they will flower continuously. While verbenas are mostly perennial plants in subtropical climates, they will not tolerate the frosts of northern European areas and should be wintered indoors.

### SCENTED FLOWER CREAM

❖

*Infuse scented flowers or petals, such as roses, violets and primroses, in thickened (whipping) cream. Use the strained liquid to make custards and creamy desserts, such as rich puddings or syllabubs. Whipped and hot, it makes a luscious topping for fruit pies or scones (biscuits).*

## *Viburnum*

***Caprifoliaceae*** (Honeysuckle family).

***Scent:*** Sweet, honeyed.

***Description:*** Many of the species in this group of evergreen and deciduous shrubs or trees are fragrant and nearly all are easily grown and extremely ornamental. My particular choice for a scented garden would be *V. fragrans* (syn. *V. farreri*), which has the bonus of flowering almost continuously throughout winter. Several spring-flowering varieties are also extremely fragrant and the summer-flowering evergreen species provide excellent long-lasting cut flowers for indoor use. Some of the viburnums have quaint names, such as Hobble Bush, Tangle-legs and Down-You-Go, a reference to their drooping branches, which have been known to root at their tips, creating 'caves' beloved of children. The bark and roots of some viburnums have been used for medicinal purposes and the leaves may be brewed in a tea.

*Viburnum carlesii*

***Species:*** *Viburnum fragrans* (syn. *V. farreri*) is an erect deciduous shrub, which was named for the British plant collector, Sir Reginald Farrer, who sent seed from the plants he found growing wild in China back to England in 1914. It is very hardy, with sweetly scented pinkish-white flowers and will reach 8-10ft/2.5-3meters. *V.* x *bodnantense,* raised by Lord Aberconway at the Bodnant Gardens in North Wales, is a lovely hybrid between *V. fragrans* and *V. grandiflorum.* It bears large clusters of fragrant, pink-flushed flowers through winter and will grow 8ft/2.4m tall. *V.* x *burkwoodii* (clusters of pure white flowers; 4-5ft/1.2-1.5m) has magnificent autumnal tones, as well as a delicious scent, as does *V. carlesii* (white flowers; 4ft/1.2m). The Malay *V. odoratissimum* (pure white flowers; 3m/10ft) is extremely attractive and its flowers are sweetly scented. However, for a real talking point, nothing beats *V. davidii,* which turns on a show of glossy sky-blue berries when flowering is complete. *V. opulus,* the Guelder Rose or Snowball Tree, and *V. tinus* both bloom well and will tolerate urban conditions, but they are only faintly scented. Nor are the varieties *V. plicatum* var. *tomentosum*, the Lace Cap Viburnums, although with their great hydrangea-like heads of flowers they are tempting to plant.

***Cultivation:*** Viburnums do best in a well-drained alkaline soil containing large amounts of leaf mold or humus. Although they like plenty of sun, they will also thrive in shade. A chalky soil is said to enhance the flowers' fragrance. Be careful not to let them dry out in summer. Viburnums may be propagated by seed or cuttings. In commercial cultivation, cultivars are usually grafted onto a stronger species and suckers later removed.

### SCENTED FLOWER SUGAR

This adds a new dimension to sweet dishes and is especially lovely when dusted over summer fruits still fresh with morning dew. Substitute scented flower sugar for ordinary sugar when preparing puddings or creamy desserts, or add it to whipped cream or custards.

*Combine caster (superfine) sugar with a handful of scented petals or flowers (fragrant roses, violets, lavender, honeysuckle or clove pinks are all suitable) in a cannister and seal.*

## *Viola*

VIOLET

*That which above all others yields the sweetest smell in the air is the violet . . .*

Francis Bacon

***Violaceae*** (Violet family).

***Scent:*** Faint but intrusively haunting; sweet.

*Viola* spp.

***Description:*** Violet, viola, pansies, heart's ease —all irresistible, even to the prophet Mohammed who said: 'As my religion is above all others, so is the excellence of the odour of violets above all others. It is warmth in summer and coolness in winter . . .' Known to the Greeks as one of Aphrodite's favorite flowers, violets were said to have first appeared when Orpheus sat down to rest one day on a mossy bank near a river. In the place where he set down his lute, the first violets burst forth from the soil. There were violets growing in the shady olive groves outside ancient Athens and the Greeks used a motif of the flower as a symbol of their city. The flowers were frequently used by physicians to 'moderate anger' and 'to comfort and strengthen the heart'. Pliny prescribed a liniment of violet root and vinegar for gout and spleen disorders and also advised that a chaplet of violets worn about the head would dispel the fumes of wine and prevent headaches.

An ancient Gaelic recommendation was to '. . . rub thy face with violets and goat's milk and there is not a prince in the world who will not follow thee!' Violet milk is an excellent skin tonic and cleanser and is advised by some doctors for acne sufferers, although the princes are not guaranteed! To make your own, place ½ cup of violet flowers in a china bowl and heat 18 fl oz/ 500ml of full cream milk almost to the boil, stirring constantly. Pour over the violets and infuse for one hour. Strain, bottle and keep in the

## PEACHY VIOLET ICE CREAM

Garnished with crystallized violets and fresh peach slices, this is a delicious finale to a special dinner party.

❖

*5½oz/150g caster (superfine) sugar*
*3–4 tbspns violets, chopped*
*9oz/250g canned peaches, drained*
*4 egg yolks, beaten*
*12fl oz/350ml thickened (whipping) cream*

*Mash the sugar, chopped violets, peaches and egg yolks in a saucepan. Stir over low heat, adding cream gradually, until the mixture forms a custardy coating on a metal spoon. Remove from the heat, pour into a bowl and chill until it begins to freeze. Remove from freezer and beat the mixture vigorously until smooth and creamy. Pour the mixture into metal freezer tray and freeze.*

*Violet leaves, at the entrance of spring, fried brownish and eaten with Orange or Lemon juice and Sugar is one of the most agreeable of all the herbaceous dishes.*

John Evelyn, *Acetaria* (1699)

## CRYSTALLIZED VIOLETS

❖

*5½oz/150g Parma violets, stems removed*
*1lb/500g white granulated sugar*
*3½fl oz/100ml water*

*In a non-aluminum saucepan, cook the sugar and water until the sugar dissolves. Add the violets and stir gently over high heat until the sugar crystallizes and dries around the flowers. Remove the pan from the heat. The sugar that remains will also be lightly scented and may be sprinkled over cakes or pies as a garnish.*

refrigerator. Soak cotton wool balls in the violet milk and pat onto the face twice daily.

Violets are also a nutrient-rich food and make a delightful addition to a green salad. A wine made from violet flowers was once popular and violet vinegar, made by steeping the flowers in white wine vinegar, has a delicious scent and a beautiful color. Violet leaves make an attractive base for holding savory molds or jellies and the flowers can be used as a garnish for chilled fruit soups with a dollop of sour cream. Violet tea, violet honey, violet pastilles, conserve of violets, violet marmalade and violet cakes are delectable recipes from bygone days.

***Species:*** *Viola blanda* (white/purple flowers); *V. cornuta* and *V. cornuta alba* (white and blue forms) include the lovely varieties 'Blue Carpet' (bright blue flowers), 'Lorna' (lavender-blue flowers), 'Duchess' (cream flowers tinged with palest blue) and 'Heather Bell' (pale mauve and pink flowers). Among the so-called Sweet Violets, *V. odorata*, look for the pink and blue cultivar 'Countess Shaftesbury', the dusky-pink 'Coeur d'Alsace' and the mauve 'Nellie Britton'. Historically, Sweet Violets were of great importance both for their medicinal value and also their culinary virtues. Records of their cultivation stretch back to before the birth of Christ. Particularly pretty is the Russia Violet, *V. sauvis*, which has very pale blue flowers. The native Australian violet or 'Johnny Jump-Up', *V. hederaceae* is, unfortunately, unscented.

***Cultivation:*** Violets are usually perennials. They love shade and appreciate a cool moist soil,

▶ Lavender and rosemary envelop the house in a heady mixture of scent.

which is slightly acid. Violets require shelter from cold winter winds in a garden, and like dappled light. They make a charming subject for rock gardens and are also very happy as edging plants spilling over banks, or as a rich carpet under a shady tree. Violets are propagated either by taking rooted runners in early winter or by sowing seed in autumn. Choose well-dug sandy loam for the plants and leave about 5in/13cm between each row. When summer comes, apply a mulch of leaf mold and well decayed cow manure and water well. Violets like as clean an atmosphere as possible.

## *Wisteria*

***Fabaceae*** (Pea family).

***Scent:*** Haunting, delicate.

***Description:*** Named in honor of the American botanist, Charles Wistar, who imported it from China to America in 1816, these beautiful deciduous climbers make an excellent ornamental display for a warm sheltered wall or (sturdy) pergola. Many old-fashioned elegant gardens feature an established 'wisteria walk', where the vines have been trained over a series of arches, creating a cool and fragrant retreat. Interplanting such walks with a non-fragrant but showily colored climber, such as the golden *Laburnum watereii,* would make a striking contrast. Wisterias may also be grown over trees or trained as standards, provided they are not placed near guttering or roof tiles, which they will gleefully destroy. Wisteria is extremely hardy and bears long, drooping racemes of pea-like flowers, which have a scent similar to laburnum. *Wisteria sinensis,* with its lovely mauve flowers, is very strongly scented, but it is the white form, *W. sinensis* 'Alba', which has the most powerful perfume of all. There are also several double wisterias, such as 'Violaceae Plena' and a pink one, *W. floribunda* 'Rosea'.

*Wisteria* sp.

***Species:*** *Wisteria sinensis,* Chinese wisteria, is probably the best for color (dark violet blue flowers) and has been known to climb 100ft/30meters. The Japanese *W. venusta* has delicately scented white flowers and interesting, downy leaves. *W. floribunda* 'Macrobotrys' (light purple flowers) bears the longest flower racemes of any wisteria, up to 3ft/1meter. Also pretty is the American *W. macrostachya,* which bears dense clusters of very pale mauve softly fragrant flowers.

***Cultivation:*** Wisterias are extremely hardy and should grow well in any sunny position. Deep rich soil is essential or they will not flower well. A firm hand in training the vines is also necessary for satisfactory flowering. Do not allow a vine to just ramble quickly over its support; it must be tied or wired in securely and encouraged to form spurs. Ensure the plant is given plenty of water prior to flowering and do not be shy when it comes to pruning each winter; hard cutting of shoots—to within 3 inches of old wood—will induce the stem to thicken.

# Bibliography


Chatto, B. *The Green Tapestry*. Collins, London, 1989.

Duff, G. *A Book of Potpourri*. Orbis Publishing Ltd, London, 1985.

Duff, G. *A Book of Herbs and Spices*. Merehurst Press, London, 1987.

Genders, R. *A History of Scent*. Hamish Hamilton, London, 1972.

Genders, R. *The Cottage Garden and the Old-Fashioned Flowers*. Pelham Books Ltd, London, 1969.

Genders, R. *A Book of Aromatics*. Darton Longman and Todd, London, 1977.

Genders, R. *Flowers and Herbs of the World*. Darton Longman and Todd, London, 1977.

Genders, R. *Scented Flora of the World*. Robert Hale, Great Britain, 1977.

Grieve, M. *A Modern Herbal*. Penguin Books, Great Britain, 1980.

Hayes, A. B. *Country Scents*. Night Owl Publishers, Australia, 1989.

Kleinman, K. and Slavin, S. *On Flowers*. Collins, Australia, 1989.

Kelly, F. *A Perfumed Garden*. Methuen, Australia, 1981.

Leyel, C. F. *Herbal Delights*. Faber & Faber, London, 1937.

Palgrave, F. T. (ed.) *The Golden Treasury of Poetry and Drama*. J. M. Dent & Sons Ltd, London, 1906.

Perry, F. *Flowers of the World*. Hamlyn Publishing, USA, 1972.

Rose, G. *The Romantic Garden*. Collins, Australia, 1988.

Squire, D. *The Scented Garden*. Doubleday Australia, 1989.

Sinclair Rhode, E. *A Garden of Herbs*. Dover Publications, Inc., New York, 1969.

Trueman, J. *The Romantic Story of Scent*. Aldus-Jupiter, London, 1975.

van Pelt Wilson, H. and Bell, L. *The Fragrant Year*. William Morrow & Co., New York.

Verey, R. *The Scented Garden*. Michael Joseph Ltd, Great Britain, 1981.

# Index

# Index

## OF SCIENTIFIC NAMES

Page numbers in **bold** indicate illustrations.

# *Pictorial Sources*

The photographs that appear in this book were kindly supplied by the following photographers and picture libraries:

David Young (DY)
Lorna Rose (LR)
Stirling Macoboy (SM)
Ivy Hansen (IH)
Bay Books Picture Library (BB)
Garden Picture Library (GPL)
Auscape International (AI)
S & O Matthews (SOM)
Joy Harland (JH)
Australian Picture Library (APL)
Densey Clyne (DC)
Tony Rodd (TR)

Individual photographs are acknowledged by page number and initials:

ii) DY; iii) DY; v) John Glover (GPL); vi) Jerry Harpur (AI); 1) DY; 3) BB; 4) BB; 5) DY; 7) Clive Boyrsnell (GPL); 11) DY; 12) LR; 13) SM; 15) IH; 17) LR; 18) DY; 19) SOM; 20) DY; 21) DY; 24) SM; 31) Linda Burgess (GPL); 33) DY; 34) Jerry Harpur (AI); 36) LR; 37) LR; 39) top-SM, bottom-SOM; 42-43) JH; 44) SM; 45) SM; 46) LR; 47) top and bottom, LR; 48) SM; 49) LR; 50) LR; 51) DY; 52) APL; 53) LR; 55) DC; 57) LR; 58) LR; 59) SOM; 60) BB; 61) LR; 62-63) Jerry Harpur (AI); 64) SM; 65) left-LR, right-IH; 67) DY; 68) LR; 70) LR; 71) LR; 73) DY; 74) SM; 75) LR; 76) LR; 77) LR; 78) LR; 79) LR; 80) DY; 81) DY; 82) top and bottom LR; 83) LR; 84) LR; 85) LR; 86) LR; 87) LR; 89) Jerry Harpur (AI); 90) TR; 92) SM; 93) LR; 95) AI; 97) left-LR, right SM; 99) right-Brett Gregory (AI); 100) SM; 102-103) APL; 104) DC; 106) DC; 107) SM; 109) Joanne Pavia (GPL); 111) SM; 112) LR; 113) DY; 115) SM; 116) John Glover (GPL); 117) LR; 118) LR; 119) LR; 121) DY; 123) DY; 125) SM; 127) LR; 128) LR; 129) DY; 131) DY; 132) LR; 133) DY; 134) DY; 137) DY; 138) LR; 139) LR; 140) BB; 141) DY; 142) DY; 144) SM; 145) SM; 146) DY; 148) LR; 150) LR; 151) DY; 152) LR; 155) DY; 156) LR; 158) TR; 159) DY; 160) left-TR, right-DY; 162-63) Michael Cook (APL); 164) DY; 166) left-DY, right-LR; 168) IH; 169) LR; 170) IH; 171) IH; 173) TR; 175) LR; 176) LR; 179) DY; 180) LR; 181) SM; 182) DY; 183) SM; 184) LR; 186) LR; 189) LR; 190) LR; 191) Andrew Henley (AI); 193) Steven Wooster (GPL); 194) JH.